Cardiovascular Engineering

Advances in Cardiovascular Physics

Vol. 5

Series Editor
D. N. Ghista, Hamilton, Ont.

KARGER

Basel · München · Paris · London · New York · Tokyo · Sydney

Cardiovascular Engineering

Part III: Diagnosis

Volume Editors
D. N. Ghista, Hamilton, Ont.; *E. Van Vollenhoven*, Leiden;
W. -J. Yang, Ann Arbor, Mich.; *H. Reul*, Aachen; *W. Bleifeld*,
Hamburg

96 figures and 13 tables, 1983

KARGER
Basel · München · Paris · London · New York · Tokyo · Sydney

Advances in Cardiovascular Physics

Vol. 2: Cardiograms: Theory and Applications. Eds. D. N. Ghista (Houghton, Mich.); E. Van Vollenhoven (Delft); W. -J. Yang (Ann Arbor, Mich.); H. Reul (Aachen).
XII + 160 p., 78 fig., 4 tab., 1979. ISBN 3-8055-2851-5
Vol. 3: Blood: Rheology, Hemolysis, Gas and Surface Interactions. Eds. D. N. Ghista (Houghton, Mich.); E. Van Vollenhoven (Delft); W. -J. Yang (Ann Arbor, Mich.); H. Reul (Aachen).
X + 166 p., 65 fig., 9 tab., 1979. ISBN 3-8055-2852-3
Vol. 4: Foundations of Noninvasive Cardiovascular Diagnostic Processes. Eds. D. N. Ghista (Houghton, Mich.); E. Van Vollenhoven (Delft); W. -J. Yang (Ann Arbor, Mich.) H. Reul (Aachen).
X + 182 p., 81 fig., 17 tab., 1979. ISBN 3-8055-2923-6

National Library of Medicine, Cataloging in Publication
Cardiovascular engineering
Volume editors, D. N. Ghista . . . [et al.]. – Basel, New York, Karger c 1983. 4 v. – (Advances in cardiovascular physics; v. 5)
Contents: pt. 1. Modelling – pt. 2. Monitoring – pt. 3. Diagnosis – pt. 4. Protheses, assist and artificial organs.
1. Biomedical Engineering 2. Cardiovascular System – physiology 3. Implants, Artificial 4. Monitoring, Physiologic – instrumentation I. Ghista, Dhanjoo N. II. Series
W1 AD53F v. 5/WG 102 C2665
ISBN 3-8055-3611-9 (pt. 3)

Drug Dosage

The authors and the publisher have exerted every effort to ensure that drug selection and dosage set forth in this text are in accord with current recommendations and practice at the time of publication. However, in view of ongoing research, changes in government regulations, and the constant flow of information relating to drug therapy and drug reactions, the reader is urged to check the package insert for each drug for any change in indications and dosage and for added warnings and precautions. This is particularly important when the recommended agent is a new and/or infrequently employed drug.

Typeset in Hong Kong by Asco Trade Typesetting Limited
Printed in Switzerland by Thür AG Offsetdruck, Pratteln
The technical aspects of this publication were conceived and developed up to the page proof stage by
Klaus Rohlfing, Verlag Gerhard Witzstrock GmbH, Baden-Baden, Köln, New York.
ISBN 3-8055-3611-9

Contents

Part III: Diagnosis
(Each of the four parts are separately available.)

Part I: Modelling

Part II: Monitoring

Part IV: Protheses, Assist and Artificial Organs

A Note from the Publisher

This four-part volume of "Advances in Cardiovascular Physics" began some time ago as a project planned, initiated, and carried through the galley proof stage by another publishing house.

Despite the exceptionally long publishing history of these books, S. Karger has accepted the responsibility for final production and distribution out of the conviction that such a significant amount of work from so many dedicated authors and editors must be honored. Work no doubt cited as 'in press' must eventually come to print.

Cardiovascular Engineering is therefore released with deep respect for the significance of this work and deep regret for the delay in its availability to those who might benefit from such a monumental effort.

Foreword (Part III)

In looking to the heart and the circulation, the mechanical engineer will find an ideal biological model to apply the laws of fluid dynamics and kybernetics. In earlier years, the diagnosis of cardiac and circulatory diseases was based mainly on clinical findings reflecting the consequences of the functional impairment of the heart as a pump. A better understanding and quantification of the physical basis of these alterations was achieved since the introduction of heart catheterization and angiography. With these techniques at hand, engineers and physicists were able to explain physiological events with the theory of fluid dynamics. They soon and additionally learned that, in constrast to models, special conditions and interactions, especially feedback mechanisms, between the heart, the vessels and the circulatory centers, alter the expected reactions. Particularly, the inhomogenity of biological materials made it necessary to simplify mathematical assumptions. As an example, although tremendous knowledge has been drawn from the application of physical laws to the heart, there is still some controversy, what "contractility" is and in which physical terms it should be expressed.

The evaluation of contractility parameters with high fidelity requires invasive techniques. Therefore, physicians and engineers began to investigate those reflections the contraction of the heart, which can be obtained from the surface of the body, such as the registration of the mechanical action of the apex (apexcardiography), the transmission of vibrations of the total heart to the chest wall, the uptake of magnetic waves (magnetocardiography) or the examination of heart sound by frequency analysis. Although all those attempts are not new and presently are less important in the clinical setting compared to easily applied other methods such as the measurement of pulmonary artery pressures and cardiac output by thermodilution with only minimal risk, it seems that the introduction of new and particularly rational engineering concepts and the use of computers will lead to new results. As an example, today the voice of a human individual can be identified with great security by computer analysis of frequency and amplitude. Why should we be less confident to achieve similar results with heart murmurs? Artifical valves could be an excellent model.

Part III gives insight into the physics of several moninvasive techniques for diagnosting heart diseases and should be very stimulating for the clinicians. It is particularly relevant, for a book of such a comprehensive scope as this one, that among the diagnostic indices and methods included herein are traditional as well as neo insights into the "contractility" concept, its clinical importance and application.

Walter Bleifeld, M.D.; F.I.C.A., Hamburg

Adv. cardiovasc. Phys., vol. 5 (Part III), pp. 1–13 (Karger, Basel 1983)

Cardiac Contractility: Physiological Basis in Isolated Cardiac Muscle, Limitations in Intact Heart

D. L. Brutsaert
University of Antwerp, Antwerp, Belgium

Contents

Introduction

Failure of the mechanical performance of the heart still constitutes one of the main causes of premature death in man. This is to a large extent due to insufficient understanding of the basic mechanisms of the contractile properties of the heart. For the evaluation of the mechanical function of the ventricle increasing attention has been paid to considering the heart as a muscle rather than as a pump. The purpose was to find properties of cardiac muscle or derived simplified indices thereof which would help in dissociating alterations of ventricular performance due to changes in loading conditions (Frank-Starling adaptation; heterometric autoregulation) from either positive (circulating catecholamines, changes in heart rate) or negative (myocardial or muscle failure) inotropic alterations (homeometric autoregulation; changes in contractility). Obviously, the pragmatic aim was to better distinguish a good from a bad heart. Simultaneously, a better insight in the fundamental biophysical characteristics of the cardiac muscle cell has resulted. Several contractile properties of isolated mammalian cardiac muscle are now well characterized. The many experimental, theoretical and conceptual criticisms have obtained careful attention and, except perhaps for the conceptual objections, have reasonably well been resolved for the isolated cardiac muscle (1). However, the direct application of these basic muscle concepts to the intact heart still remains controversial and can be criticized on several grounds, e.g. the difficult experimental measurements, the many oversimplified assumptions in modelling the ventricle, insufficient understanding of basic cardiac mechanics, failure to take into consideration some conceptual limitations, the complex physiological loading of a muscle-pump system, etc. Recent experimental work has permitted a coherent and unifying description of heart muscle performance. The basis of it is that the physiological principles governing mechanical activity of myocardial muscle can be defined in terms of the interrelationship between force development, instantaneous length during shortening, and velocity of shortening, independently of the time course of activation. This approach of cardiac muscle mechanics requires neither extrapolations nor assumptions based on models. This analysis merely provides an objective description of what the muscle really does and how it behaves when its loading conditions are altered. The first part of this chapter summarizes some of the experiments which have led to this description. It also provides a critical basis for defining the concepts of contractility and changes in contractility in the isolated cardiac muscle. In the second part, the limitations of these muscle concepts in the intact ventricle will be dealt with. Pragmatic usefulness as the only justification is no longer tenable as a sufficient argument for such applications in this important and controversial field of medicine. Emphasis will be placed upon the need for reintegrating muscle and pump function. The ventricle is primarily a combined muscle-pump system and neither a papillary muscle nor even less a skeletal muscle. Muscle concepts as directly derived from skeletal muscle physiology should therefore no longer forcibly be applied to the intact heart since this practice may perhaps explain part of the stagnation in the field of ventricular function evaluation over the past few years. Rather, research should be made into these properties of heart muscle function which would help to define its behaviour while simultaneously integrated in the wall of a pumping system. Obviously, such considerations can at present only provide directional information rather than clearcut answers, new socalled indices etc. These principles have recently been published in two extensive review articles (*1a*, *1b*). In the present chapter only the most essential points will be recapitulated. The interested reader will find a more complete and updated literature analysis in the above mentioned reviews.

A. The Force-Velocity-Length-Time Relationship of Isolated Cardiac Muscle

There is general agreement that a complete analysis of the mechanical performance of a twitch contraction of mammalian heart muscle requires consideration of at least four variables, i.e. force development, length during shortening, velocity of shortening and the time after the onset of activation (Fig. 1). Although the fundamental force-length and the inverse force-velocity relationships of tetanized skeletal muscle have been applied to twitch contractions of heart muscle (2, 3, 4), the analysis of the latter relationship is complicated by the con-

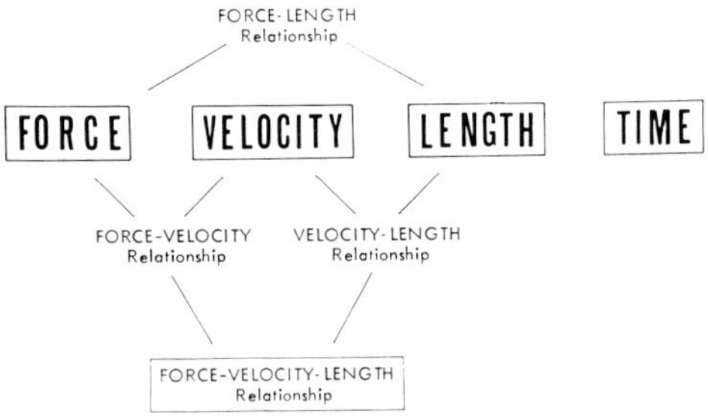

Fig. 1 Principal determinants of a twitch contraction in heart muscle. The force-length and force-velocity relationships are fundamental properties derived from tetanized skeletal muscle. The velocity-length relationship has permitted a new experimental approach for the analysis of force-velocity-length-time interrelationship of twitch contractions in isolated cardiac muscle.

tinuously changing length and time during a twitch (5, 6), while the former one lacks information about velocity of shortening and time.

Quick release techniques have been used in literature in order to correct for length and time to obtain instantaneous force-velocity relations during twitch contractions (7, 8, 9). However, much confusion in heart muscle mechanics has arisen from the misinterpretation of these techniques. As outlined elsewhere (1), several inherent problems, such as the late time and short lengths at which the post release velocity is measured, along with the model dependence, inactivation (10, 11) and the need to readjust the damping of the system for releases resulting in force steps of varying amplitudes in a non linear elastic system such as a muscle, would indicate that these techniques should be abandonned. Quick release techniques are inappropriate for physiological or pharmacological analyses of cardiac muscle performance at longer lengths, e.g. within 10% below l_{max}, and in the early phase of the contraction.

Recently, various new experimental techniques, e.g. phase-plane velocity-length analysis (11), load clamping (12) and zero load clamping (13, 14, 15), were developed (Fig. 2). These techniques have permitted the simultaneous analysis of the four variables, force-velocity-length-time, in twitch contractions, and have resulted in a new appreciation of the way in which velocity is influenced by the changing length, load and time.

The phase-plane analysis of velocity of shortening as a function of instantaneous length during shortening at a given load is illustrated in Fig. 2A for a preloaded and an afterloaded isotonic contraction. When these phase-plane velocity-length curves are recorded for increasing afterloads and plotted in a threedimensional way, a surface is described characterizing active heart muscle during external shortening in terms of velocity, length and force (Fig. 3).

Through the application of a second new experimental technique (Fig. 2B), in which sudden load alterations (load clamp) (12) are imposed on the shortening muscle, it has become clear that the variables force, velocity and length are intimately interrelated and independent of the time course of activation over a large portion of the contraction. This time independence extends over a major portion of external shortening, its duration being partly determined 1. by the absolute loading conditions and 2. by the socalled instantaneous uncoupling effect of the preceding shortening or length changes per se during that same contraction (1).

The analysis of this force-velocity-length interrelationship has been further extended recently to loads smaller than the preload by the development of a third new experimental technique (Fig. 2C) whereby the muscle is unloaded during shortening immediately following the latent period from the preload to zero external load (zero load clamp) (15) or to intermediate loads. In this way the force-velocity-length relationship can be studied over the entire loadrange from peak isometric force to close to zero load and over the entire physiological length range below l_{max}. Moreover, through the application of early zero load clamps the onset of mechanical activation can be analyzed indicating an early onset of the time independence (14, 15).

For any constant total load, such as during an isotonic contraction, active heart muscle moves over the force-velocity-length surface in a predictable fashion during external shortening (Fig. 3, hypothetical contraction in thick line). Velocity is only determined by the instantaneous length, regardless of the initial muscle length from where the contraction started shortening, – and hence of the preload

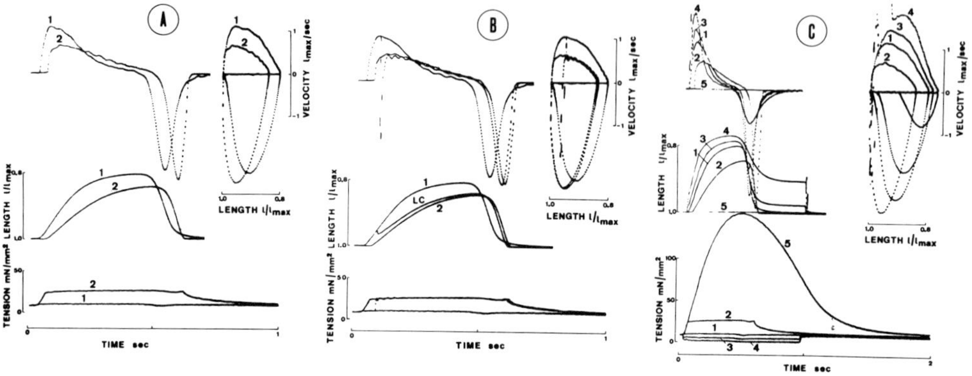

Fig. 2 New experimental techniques for the study of twitch contractions of heart muscle.
A. Phase-plane velocity-length relationship:
Contractions 1 and 2 are isotonic twitch contractions with respectively preload only and preload plus afterload.
Left: shortening velocity, shortening and tension are displayed as functions of time.
Right: phase-plane velocity-length relationship.
B. Load clamp:
Contractions 1 and 2: same as in A.
LC: contraction in which the load is abruptly augmented from the loading conditions of contraction 1 to those of contraction 2. Following the clamp step the phase-plane velocity-length relation of LC coincides with the velocity-length relation of contraction 2 during a major part of external shortening despite a dissociation of the time course of the length changes (left, middle) of both contractions, indicating time independence of the force-velocity-length relationship.
C. Zero load clamp:
Contractions 1 and 2: same as in A and B.
In contraction 4 the load (preload) was suddenly removed following the latency period to zero external load (zero load clamp). Damping of the entire muscle-lever system was critically adjusted. In contraction 3 an intermediate load between the preload and zero external load was imposed. The first isometric beat after a series of stable contractions at the preload (contraction 1), is indicated by the number 5.
Cat papillary muscle; cross-sectional area 0.56 mm^2; all contractions were recorded in the physiological length range with an initial muscle length at l_{max} (8 mm), i.e. the length were active tension development was maximal; 29° C; 12/min; recordings on storage display unit (Tektronix 611) with hard copy unit (Tektronix 4601). Notice change in time scale in C.

within the length range below l_{max}, – and regardless of the time after activation.

These properties were extensively discussed in two recent review articles (1, 16) in which the interested reader will find a substantive discussion of the influence of changing length and time on this force-velocity-length interrelationship, of the possible underlying mechanisms, and of the implications and limitations.

Before proceeding with the possible application of this description to the intact heart let us first consider the concept of "contractility" or "contractile behaviour or state" of heart muscle. Contractility is a concept which is frequently alluded to in cardiac physiology with regard to the performance of the heart (4, 17). However, as rightly pointed out by several investigators (10, 18, 19) this term has had no agreed and precise definition. This confusion is partly due to the many controversies in cardiac literature, the major reason being that no simple unifying description has been given so far of heart muscle performance. Rather incomplete descriptive definitions have been advanced previously by Leonard and Hajdu (17) who define "changes in contractility" as a change in the "force" development independent of changes in initial length, and by Sonnenblick (6) and Blinks and Koch-Weser (4) who also include changes in active state

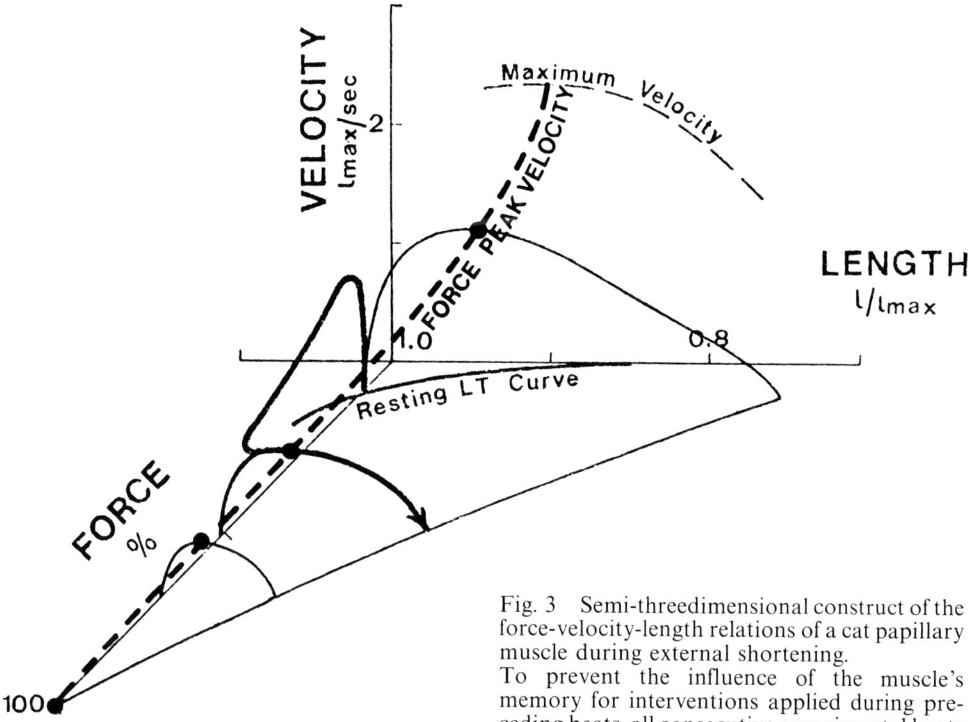

Fig. 3 Semi-threedimensional construct of the force-velocity-length relations of a cat papillary muscle during external shortening.
To prevent the influence of the muscle's memory for interventions applied during preceding beats, all consecutive experimental beats have been separated by a constant number of intervening equally loaded control beats, e.g. here 5 to 10 isotonic contractions with the preload maintained throughout (contraction 1 of Fig. 2). The phase-plane velocity-length relations of isotonic contractions obtained at l_{max} have been replotted as a function of total load. The phase-plane curves can either indirectly be redrawn or directly displayed on a storage display unit (Tektronix 611) in a semi-threedimensional way by subtracting force from the length and from the velocity signal by means of two operational amplifiers. The phase of relaxation has been omitted. The threedimensional construct is defined by three planes: the velocity-length plane to the right, the force-velocity plane to the left, and the basal plane represents the isotonic force-length relations. The dashed line in the velocity-length plane indicates part of the velocity-length relations of a zero load clamped beat (see also Fig. 2C). Maximum unloaded shortening velocity at zero external load remains constant up to 12.5% below l_{max}. The initial part during the zero load clamp (not displayed on the figure) is probably model dependent. Maximum unloaded shortening velocity becomes markedly length dependent in the shorter (>12.5% below l_{max}) length range.
Resting length-tension (LT) curve is indicated on the basal plane. The force-peak velocity curve depicts one single curve for the overall force-velocity-length surface below l_{max} and merely provides a useful representative profile through this surface. As outlined elsewhere, both for practical and theoretical reasons, this single two-dimensional force-peak velocity curve derived from a series of contractions at the single resting length, l_{max}, can sufficiently and adequately characterize cardiac muscle's mechanical capabilities 1. within the physiological length range (between l_{max} and approximately 12.5% below l_{max}), where as noted maximum unloaded shortening velocity is constant with length, 2. after activation has fully been achieved, and 3. over the full range from isometric to near zero force. The velocity course of a hypothetical afterloaded contraction is superimposed (thick line). The velocity of shortening during the isometric phase of this contraction has been theoretically derived from a two-component model for didactic purposes $\left(V_{CE} = \frac{dP/dt}{kP + C}\right)$.
The larger isotonic part of the force-velocity-length surface during external shortening including the force-peak velocity curve and the plateau of maximum unloaded shortening velocity is not model dependent.

and the force-velocity relation. Sonnenblick (6) and Brady (5) first stressed the importance to also consider instantaneous length. The definitions of static and dynamic contractility, as proposed by Brady (10), are entirely interpretative "model" dependent. Contractility can now be redefined in view of the present knowledge of cardiac muscle performance (Fig. 4). The distinction between the terms "contractility" and "change in contractility" is essential for this purpose. Contractility can be defined as the "time independent interrelationship between force velocity and length". In this sense, changes in initial muscle length – or preload – do not influence contractility. The term "changes in contractility" then reflects any level shift in this time independent interrelationship, i.e. when for any given constant total load and for any instantaneous length the velocity of shortening is altered. This moves the entire surface into a new position, up (calcium, catecholamines, stimulation frequency) or down (myocardial failure, anoxia) which is again unique for this new contractile state independent of initial muscle length (preload) and of the time after activation.

B. Problems in the Intact Heart

While these basic principles may be clear from the analysis of isolated cardiac muscle mechanics, the direct translation to the intact heart is still limited. At least five important problems must be taken into consideration before these concepts can be usefully applied to the intact heart in attempts to exclude the effects of "volume and loading" (contractility) from considerations of "changed contractile state" (changed contractility).

1. Loading and Length History of Preceding Contractions

A first problem can be derived from the fact that loading (20, 21) and length (22) history of

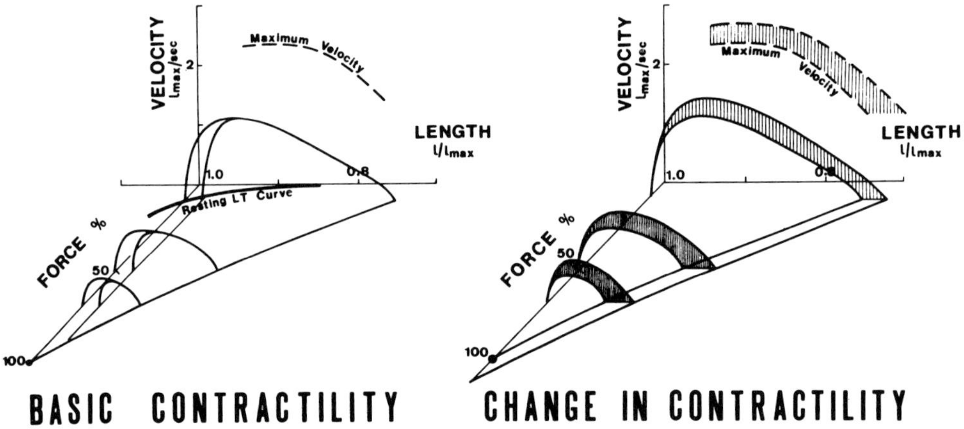

Fig. 4 Basic contractility is to be defined as the "time-independent part of the force-velocity-length relation". This part encompasses most of the force-velocity-length surface during external shortening except near peak shortening, and is independent of the initial muscle length from where the muscle starts shortening. The phase of relaxation is not included in this definition. Part of maximum unloaded shortening velocity, as obtained by zero load clamping the muscle immediately following the latency period, is indicated as a dashed line in the velocity-length plane. The midportion of this maximum velocity in the length range between 5 and 12.5 % below l_{max} can be regarded as a potentially useful index of "contractility" 1. being measured at nearly zero load, i.e. zero external load plus the critical damping force, 2. being constant (i.e. within 5 % of maximum) with length in this physiological length range, and 3. after full activation at zero load, and hence time independence, has been reached. The first part of maximum unloaded shortening-velocity, i.e. > 95 % l_{max}, during the zero load clamp being model dependent, and the terminal part in the shorter length range below 87.5 % l_{max} being length dependent, cannot be used for this purpose. Change in contractility: shift of the time independent part of the force-velocity-length relation with concomitant shift of maximum unloaded shortening velocity in the length range between 95 and 87.5 % l_{max}.

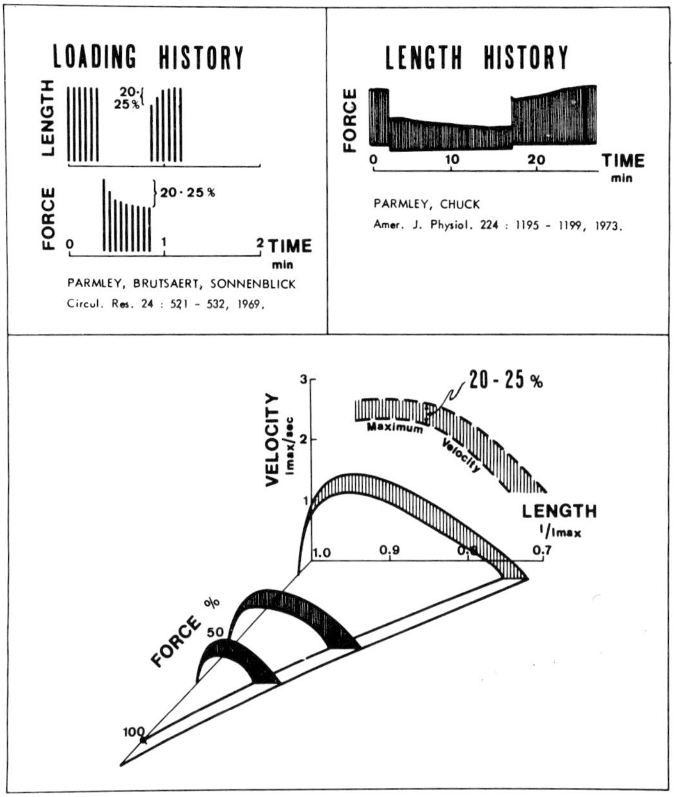

Fig. 5 Upper left: Sudden change from isotonic contractions to isometric and back to isotonic. Due to the slow time scale, each contraction depicts a single vertical line. The first isometric beat is 20 to 25 % higher than the subsequent stable isometric beats, and alternatively the first isotonic beat is 20 to 25 % lower than the subsequent stable isotonic beats. Isometric d*P*/d*t* and isotonic d*l*/d*t* undergo directionally similar changes. Time to peak isometric force of the first isometric beat is also slightly longer than in the subsequent beats.
Upper right: Sudden change of initial muscle length from a higher to a lower value. The slower time scale indicates that the concomitant changes in performance occur much slower than when the loading conditions are altered.
Lower: Force-velocity-length relations depicting shifts of the entire surface due to changes in loading and/or length conditions of the preceding contractions. These shifts cannot be distinguished from shifts due to true changes in contractility (Fig. 4, right).

the preceding contractions are important determinants of cardiac muscle performance (Fig. 5). When the loading (20, 21) of a series of contractions is suddenly altered, e.g. from isotonic to afterloaded isotonic or to isometric or alternatively when the initial muscle length (22) is abruptly changed, important time-dependent apparent changes in contractility with a concomitant shift of the entire surface (23) are noted. Accordingly, unless these variables are rigorously controlled as in all our isolated papillary muscle experiments, one cannot simply relate velocity to length and load without a clear knowledge of the previous history of the muscle. In order to correct for this long term memory following any loading or initial length alteration, it is not enough to allow the muscle for stabilisation to a socalled steady state contraction. All experimental beats must be separated by a stabilized series control beats at a preset well determined loading and length condition, i.e. all experimental beats must refer to the same original force-velocity-length surface. Any experimental situation as in the intact heart where these variables cannot be rigorously held constant, is therefore open for criticism, for there is no way to distinguish shifts of the force-velocity-length surface due to a true change in contractility from the influence of this length and load dependent memory of the muscle.

This memory of long duration for loading and length history of preceding contractions should not be confused with the direct and instantaneous effects of absolute load and/or length changes observed during the course of a same contraction. This latter short term memory of the muscle is an inherent property of both skeletal (24, 25, 26, 27, 28, 29) and cardiac (10, 11, 29, 30) muscle and has extensively been discussed previously (1).

2. Model Dependence

A second problem deals with the fact that the application of muscle concepts to the intact heart depends on the assumption of analogue mechanical models. However, the most appropriate mechanical model, Voigt vs Maxwell, that best represents actual performance is still unclear (10, 31, 32). There also still remains some controversy as to whether the stiffness of the series elastic elements (SE) is only force dependent (31, 32, 33), – and hence essentially fixed in time as in the classical Hill model (34) – or time dependent (35, 36), – and hence continuously changing with the number of cross-bridges activated in parallel as in the Huxley-Simmons model (37) – , or to a combination of both (38). These considerations become important when during the isovolumic phase of systole length corrections are made for internal shortening due to SE extension or when contractile velocity element (V_{CE}) is calculated.

Another point of concern is the nature of the parallel elastic elements (PE). Evidence from skeletal muscle fibers (39, 40) and from cardiac muscle (41, 41a) would indicate that this significant restoring force in heart muscle would be rather active than passive.

The sensitivity of shortening velocity of cardiac muscle to total load (12) could also be in line with this view. Hence, both nature (active vs passive) and/or localization (42) of the PE will determine any extrapolation from V_{CE} values to zero load in order to obtain V_{max} as an index for contractility in the intact heart.

3. Indices of Contractility and Changes in Contractility

A threedimensional construct of force-velocity-length relations is difficult to obtain in the intact heart and would be far too complex to be useful as a quantitative measure for the analysis of afterloaded beats in the physiologically relevant midportion of the surface. What "indices" are known to characterize the surface in a more simple fashion? Ideally, such indices should reflect either the three variables, force-velocity-length, simultaneously and in the time independent part of the contraction, or one of the three variables independently of the two other ones and also in the time independent portion. An index, fulfilling the former criteria has still not been derived. In the isolated cardiac muscle (Fig. 4) maximum unloaded velocity of shortening (V_{max}) could represent an index which satisfies the latter criteria, being:
measured at zero load,
constant with length in the longer physiological length range (87.5–95% l_{max}) (1, 13),
independent of time after 15–20% of the time to peak tension (14, 15, 16) and
sensitive to changes in contractility (43).
Through the application of zero load clamp techniques, permitting direct measurements of maximum unloaded shortening velocity at zero external load, and through phase plane analysis of such measurements, most experimental and theoretical criticisms against the use of V_{max} as an index of contractility in isolated cardiac muscle have been resolved. Its use as an index is limited however
by the length dependence in the shorter (<87.5% l_{max}) length range (Fig. 4) (1, 13), and
by the time dependence at longer lengths
a) after the addition of caffeine in mammalian heart muscle (14),
b) after the substitution of strontium for calcium in mammalian heart muscle (15), and
c) in frog heart muscle (16, 44),
by the conceptual criticisms (see below).

Nevertheless, despite the unquestionable usefulness of directly measured maximum unloaded shortening velocity as an index of contractility in isolated cardiac muscle, indirect derivations of V_{max} as an exclusive index of contractility in the intact heart must be viewed with caution due to the many oversimplifications in modelling and extrapolations from calculated force-velocity measurements. What other indices can be derived from the application of muscle mechanics to the intact heart? In order to be useful for the evaluation of contractility any direct or derived hemodynamic measurement of pressures and volumes during the pre-ejection and during the ejection phase must reflect the simultaneous interrelation between the three variables, force, velocity and length, during the time independent portion of the contraction. None of all the presently proposed measurements has satisfied this criterion (45). Many hemodynamic measurements reflect one or two of these variables, but no single measurement encompasses all three of them. This is also the case for non invasive measurements and indices. In addition to this important limitation, one of the most widely used derived indices, V_{CE}, is open for other severe criticisms. The lack of an appropriate analogue mechanical model which is necessary for the calculation of V_{CE} has already been discussed. Since isolated cardiac muscle always feels total load and has no memory for its preloading or afterloading condition within the time independent portion of a given contraction, any analysis in the intact heart which considers developed rather than total pressure is based on a misconception of basic muscle mechanics. The model dependence of k and C and the possible alterations with various disease states may also complicate these calculations.

Furthermore, whereas during zero load clamp experiments in isolated cardiac muscle one can still argue about the conceptual basis of extrapolations from measured maximum velocity values to the theoretical V_{max} over less than 1% of peak isometric force, extrapolations in the intact heart over more than 10 to 20% must be considered with caution. Peak V_{CE} and peak V_{CF}, during preejection and ejection phases respectively, represent only single points somewhere on the force-velocity-length surface, are thus length and load dependent, and hence can merely be used as pragmatic arbitrary tools for ventricular function evaluation. Moreover, the impact of asynchrony and temporal or spatial inhomogeneity of ventricular contraction is still insufficiently understood, to permit oversimplifications. The ventricle is not a single papillary muscle. Both segmental and transmural differences of contractile behaviour occur, and may limit the application of a single and unique concept of contractility. For that matter new concepts, such as segmental and transmural contractility, were recently introduced (45).

4. Conceptual Criticisms

Along with the development of the concept of a time independent force-velocity-length relationship, the many experimental, theoretical and conceptual criticisms that were formulated against studies on cardiac muscle mechanics have received careful attention and, except for the conceptual objections, have been resolved reasonably well (1).

a) Dual system of calcium control in cardiac muscle?

Both in isolated cardiac muscle and in the intact heart, the conceptual basis of 'changing contractility' still represents an unresolved major criticism which warrants careful consideration. As contraction is caused by the cyclic attachment and detachment of myosin-crossbridges of the thick filament with the action of the thin filament, activating calcium, in principle, could modify performance by affecting regulatory systems on either the thick or the thin filaments. The troponin-tropomyosin based thin-filament-linked regulation could act as an on-off switch; a thick-filament-linked system would, instead, vary performance by modulating the actin-myosin interaction once the thin filament is switched on. Yet, is there sufficient evidence for a dual system of calcium control involving both the thin and the thick filaments in cardiac muscle? Does a shift in the force-velocity-length surface, with a concomitant shift in V_{max}, reflect a true alteration in the actinomyosin ATPase turnover rate of each individual crossbridge between the actin and myosin myofilaments (thick-filament-linked regulation) or does it merely reflect a change in the number and rate of crossbridge interactions (thin-filament-linked regulation) (19)? In other words, does there really exist a 'change in contractility' which would in any way be different from a change in force potential of the muscle? If this were not the case, then this would mean that alterations in measured maximal unloaded shortening velocity at zero external load, e.g. when the external calcium contraction is changed, might simply reflect the presence of an unrecognized internal load due to intracellular friction or a too slow rate of detachment of crossbridges (43, 46). This load would be carried by a changed number of activated sites so that the load per active site would be altered, and not by a changed intrinsic speed of the enzymatic reaction at each site. Investigations on skinned skeletal muscle fibres (40), showing that calcium sensitivity of V_{max} is greatest at lower ionic strength and disappears at higher ionic strength, would be in favour of this view. Although the control mechanisms of actin and myosin interaction may be different in cardiac and skeletal muscle, experiments in isolated electrically stimulated cat papillary muscles (47) and in isolated adult rat heart cells (47a) are only partially consistent with these observations. When the osmolality of the external medium was raised by adding sucrose or

NaCl, the sensitivity of V_{max} for calcium appeared essentially unaltered in the lower calcium range, but it became less sensitive, although in contrast to skeletal muscle not totally insensitive, to calcium changes in the higher calcium range. Moreover, the effect of ionic strength on skinned fibres (40) could not be confirmed in intact skeletal muscle (47b). Moreover, more recent experimental evidence also suggests an additional calcium-dependent thick filament-linked regulation in vertebrate striated muscle (47c, 47d, 47e). In the meantime, as long as this controversy is not resolved this conceptual criticism still represents the most important obstacle against the useful application of muscle mechanics in ventricular function evaluation, for this would mean that the distinction between the effects of volume and loading (basic contractility) and true alterations in contractility, as being reflected either in force potential or in velocity parameters respectively, would merely be an illusion.

b) Control of cardiac performance by length or by changed contractility?

Jewell (47f) in an excellent review provided evidence that length control of cardiac muscle performance in the ascending portion of the Starling curve would be governed by alterations in calcium activation rather than by the degree of overlap or steric hindrance of the myofilaments... 'if the inotropic state of the muscle is equated with the degree of activation of the contractile system, then muscle length influences inotropic state and a change of muscle length must therefore be regarded as an inotropic intervention; if these results are applicable to the intact heart, then diastolic volume and inotropic state cannot be regarded as independent regulators of cardiac output'. Still, although this problem seems far from resolved, from a practical point of view the striking differences in the time constants by which length and changes in contractility control cardiac performance might necessitate to further consider both mechanisms separately.

5. Physiological Loading

Ever since the basic principles of cardiac muscle mechanics, as derived from classical skeletal muscle concepts, have been applied to the intact heart, it has been widely stated in almost any publication or textbook, that the evaluation of ventricular performance as a muscle should include at least three important determinants i.e. the preload, the afterload and changes in contractility. The distinction between preload and afterload has undoubtedly been useful for an easy extrapolation from a muscle system into a pump system, the preload being the initial volume or fiber length of the ventricle as determined by the filling pressure, and the afterload being the resistance to ventricular emptying as determined in a complex manner by the aortic pressure.

However, this distinction how fundamental and useful it might be in isolated muscle systems, is another typical example of how basic muscle concepts have forcibly and misleadingly been applied to the intact heart. It also neglects the intimate and continuous interrelationship between force, length and velocity. The total loading represents the only physiological determinant of a contracting heart muscle-pump system for the following reasons:

1) A given state of contractility of cardiac muscle can be characterized by a single force-velocity-length surface. Within physiological limits, alterations both in preload or initial fiber length and in afterload simply signify displacements of cardiac muscle contraction on that same surface. Moreover, in a given beat velocity of shortening at a given length is only determined by the total load (12) and has no memory, at least not during the time independent part of the contraction and insofar the memory of the muscle for the loading conditions of the preceding beats (20, 21) has been taken into consideration, for its immediate and separate preloading and afterloading condition.

2) In addition to this fundamental aspect that cardiac muscle always senses total load, in the intact heart there is a continuous interplay between preloading and afterloading conditions. Only during acute extreme volume overloading, when all the sarcomeres across the wall of the normal ventricle become maximally extended, the ventricle can be considered to operate as if it were at a fixed end – diastolic volume and hence constant preload. Any further increases in aortic pressure and volume then only serve to augment the afterload (48). Hence, in general, preload and afterload can hardly be distinguished and their mutual influence can hardly be recognized except under specific conditions. The distinction between a disease state which is primarily due to a preload alteration, e.g. volume over-

loading, or which is primarily due to an afterload alteration, e.g. pressure overloading, can be useful from a didactic point of view or in order to predict or explain the ensuing adaptation (dilatation, hypertrophy) in the chronic condition, but it is very misleading in terms of muscle mechanics extrapolated to the intact heart, e.g. how does a muscle fiber in the right ventricular wall feel a large left to right shunt in terms of preload or afterload? Or similarly, what are the eventual implications of a valvular insufficiency or stenosis in left or right ventricle? Or how can the additional load due to the inertial and Bernouilli-effects (see below 3) of an augmented end – diastolic volume be distinguished from the true static afterload? The only functional implication of such alterations for a ventricular wall muscle fiber is the total loading condition at any instant during ventricular contraction. Besides, the main aim for the application of muscle mechanics to the intact heart for the evaluation of the basic performance capacity of this pump has been to determine a parameter or a description of ventricular performance that is independent of its total loading condition (Frank-Starling adaptation), regardless of whether this is determined primarily by diastolic volume or by aortic pressure, and is thus a measure of contractility or changes in contractility.

3) Ventricular load has incompletely been described as the wall stress resulting from a complex function of intraventricular pressure and ventricular geometry. Only classical loading concepts (preload and afterload) as encountered in muscle systems extrapolated to a circumferential wall fiber in the ventricle and recalculated to wall stress applying the Laplace relationship were taken into account. Except for a few recent studies (49, 50), too little attention has been paid to those aspects of cardiac muscle performance which are related to its physiological integration in a pumping system (51, 52). Abbott and Gordon (49) have recently proposed to also consider additional loads which resulted from a hydrodynamic analysis of left ventricular pumping. Physiological loading on a muscle fiber in the ventricular wall represents an extremely complex function (49, 53)

$$f_m = Af_0 + Bv + Ca + Dv^2 + \ldots$$

where

f_0, static loading (preload and afterload),
v, velocity of shortening of the muscle,
a, acceleration of the muscle,
A, B, C, D are constants slowly varying with time due to the Laplace relationship.

f_m thus consists of a static load (preload and afterload) and a summation of complex dynamic loading functions, each of which being derived from the pumping action of a ventricle:

a. a load which is proportional to the velocity of shortening (v) accounting for the different damping effects on the contracting cardiac muscle;

b. a load proportional to the acceleration of shortening (dv/dt or a) accounting for inertia, i.e. moving mass of the blood, the mass of the recoiling vessel walls, the mass of the contracting walls of the heart, etc ...;

c. a load proportional to the square of shortening velocity (v^2) accounting for the load resulting from the geometrical constriction of the outflow tract of the ventricle (Bernouilli-effect).

This equation should include many more other functions as wall, i.e. some aspects of the impedance of the vascular system (50), lateral compressive forces on the muscle by the surrounding fibers during systole (54), mutual influences of these various loads on each other, etc. (55)

Accordingly, the physiological loading of a muscle fiber in the ventricular wall can hardly be calculated being an extremely complex function continuously changing with time. Recently, similar varying loads were imposed on contracting cat papillary muscles (53). As long as the loading conditions were changing shortening velocity at any length and total load deviated from the previously defined time independent force-velocity-length relation derived from cardiac muscle experiments under static loading (constant preload + afterload) conditions. Shortening velocity was lower at any length and total load, when loading was dynamically increasing and was higher as long as the muscle was being dynamically unloaded. Again, there is no way to distinguish these conditions of altered shortening velocity at any length and total load from true alterations in contractility. Accordingly, the application of this concept of contractility or of any derived index resulting from statically loaded isolated cardiac muscle experiments becomes quite unreliable for the interpretation of data derived under dynamically changing loads as encountered in a pumping ventricle.

References

1. Brutsaert D L: The force-velocity-length-time interrelation of cardiac muscle. In: The Physiological Basis of Starling's Law of the Heart. Ciba Foundation, London 1974, pp 155–175.

1a Brutsaert D L, Paulus W J: Loading and performance of the heart as muscle and pump, Cardiovasc Res 11, 1–16 (1977)

1b Brutsaert D L, Paulus W J: Contraction and relaxation of the heart as muscle and pump. Internat Rev Physiol 18, 1–31 (1978)

2. Abbott B C, Mommaerts W F H M: A study of inotropic mechanisms in the papillary muscle preparation. J Gen Physiol 42, 533–551 (1959)

3. Sonnenblick E H: Implications of muscle mechanics in the heart. Fed Proc 21, 975–990 (1962)

4. Blinks J R, Koch-Weser J: Physical factors in the analysis of the actions of drugs on myocardial contractility. Pharmacol Rev 15, 531–590 (1963)

5. Brady A J: Time and displacement dependence of cardiac contractility: problems in defining the active state and force-velocity relations. Fed Proc 24, 1410–1420 (1965)

6. Sonnenblick E H: Determinants of active state in heart muscle: force, velocity, instantaneous muscle length, time. Fed Proc 24, 1396–1409 (1965)

7. Sonnenblick E H: Active state in heart muscle: its delayed onset and modification by inotropic agents. J Gen Physiol 50, 661–676 (1967)

8. Edman K A P, Nilsson E: Mechanical parameters of myocardial contraction studied at a constant length of the contractile element. Acta Physiol Scand 72, 205–219 (1968)

9. Noble M I M, Bowen T E, Hefner L L: Force-velocity relationship of cat cardiac muscle, studied by isotonic and quick-release technique. Cir Res 24, 821–833 (1969)

10. Brady A J: Active state in cardiac muscle. Physiol Rev 48, 570–600 (1968)

11. Brutsaert D L, Sonnenblick E H: Force-velocity-length-time relations of the contractile elements in heart muscle of the cat. Circ Res 24, 137–149 (1969)

12. Brutsaert D L, Claes V A, Sonnenblick E H: Effects of abrupt load alterations on force-velocity-length and time relations during isotonic contractions of heart muscle: load clamping. J Physiol (London) 216, 319–330 (1971)

13. Brutsaert D L, Claes V A, Sonnenblick E H: Velocity of shortening of unloaded heart muscle and the length-tension relation. Circ Res 29, 63–75 (1971)

14. Henderson A H, Claes V A, Brutsaert D L: Onset of unloaded velocity of shortening in mammalian heart muscle as influenced by caffeine and other inotropic interventions: time course of activation. Cir Res 33, 291–302 (1973)

15. Brutsaert D L, Claes V A: Onset of mechanical activation of mammalian heart muscle in Calcium- and Strontium-containing solutions. Circ Res 35, 345–357 (1974)

16. Brutsaert D L, Henderson A H: Time course of mechanical activation in cardiac muscle. Eur J Cardiol 1, 201–208 (1973)

17. Leonard E, Hajdu S: Action of electrolytes and drugs on the contractile mechanism of the cardiac muscle cell. In Handbook of Physiology, Circulation, vol 1. American Physiological Society, Washington 1962, pp 151–197

18. Blinks J R: On the measurement of myocardial contractility. Anesthesiology 28, 800–803 (1967)

19. Katz A M: Contractile proteins of the heart. Physiol Rev 50, 63–158 (1970)

20. Parmley W W, Brutsaert D L, Sonnenblick E H: Effects of altered loading on contractile events in isolated cat papillary muscle. Circ Res 24, 521–532 (1969)

21. Kaufmann R L, Lab M J, Hennekes R, Krause H: Feedback interaction of mechanical and electrical events in the isolated mammalian ventricular myocardium. Pflügers Arch, Eur J Physiol 324, 100–123 (1971)

22. Parmley W W, Chuck L: Length-dependent changes in myocardial contractile state. Am J Physiol 224, 1195–1199 (1973)

23. Henderson A H, Ocken E Van, Brutsaert D L: A reappraisal of force-velocity measurements in heart muscle. Eur J Cardiol 1, 105–118 (1973)

24. Jewell B R, Wilkie D R: An analysis of the mechanical components in frog's striated muscle. J Physiol (London) 143, 515–540 (1958)

25. Jewell B R, Wilkie D R: The mechanical properties of relaxing muscle. J Physiol (London) 152, 30–47 (1960)

26. Gordon A M, Huxley A F, Julian F J: The variation in isometric tension with sarcomere length in vertebrate muscle fibres. J Physiol (London) 184, 170–192 (1966)

27. Hill A V: First and Last Experiments in Muscle Mechanics. Cambridge University Press, London 1970

28. Taylor S R, Rüdel R: Striated muscle fibres: inactivation of contraction induced by shortening. Science 167, 882–884 (1970)

29. Edman K A P, Nilsson E: Time course of the active state in relation to muscle length and movement: a comparative study on skeletal muscle and myocardium. Cardiovasc Res 5, Suppl 1, 3–10 (1971)

30. Brutsaert D L, Claes V A, Donders J J A: Effects of controlling the velocity of shortening on force-velocity-length and time relations in cat papillary muscle: velocity clamping. Circ Res 30, 310–315 (1972)

31. Parmley W W, Sonnenblick E H: Series elasticity in heart muscle: its relation to contractile element velocity and proposed muscle models. Circ Res 20, 112–123 (1967)

32. Hefner L L, Bowen T E: Elastic components of cat papillary muscle. Am J Physiol 212, 1221–1227 (1967)
33. McLaughlin R, Sonnenblick E H: Time behavior of series elasticity in cardiac muscle: Real-time measurements by controlled length techniques. Circ Res 34, 798–811 (1974)
34. Hill A V: Heat of shortening and the dynamic constants of muscle. Proc R Soc Lond (Biol) 126, 136–195 (1938)
35. Noble M I M, Else W: Reexamination of the applicability of the Hill model of muscle to cat myocardium. Circ Res 31, 580–589 (1972)
36. Pollack G H, Huntsman L L, Verdugo P: Cardiac muscle models: An overextension of series elasticity. Circ Res 31, 569–579 (1972)
37. Huxley A F, Simmons R M: Proposed mechanism of force generation in striated muscle. Nature (Lond) 233, 533–538 (1971)
38. Meiss R A, Sonnenblick E H: Dynamic elasticity of cardiac muscle as measured by controlled length changes. Am J Physiol 226, 1370–1381 (1974)
39. Hill D K: Tension due to interaction between the sliding filaments in resting striated muscle: effect of stimulation. J Physiol (London) 199, 637–684 (1968)
40. Thames M D, Teichholz L E, Podolsky R J: ionic strength and the contraction kinetics of skinned muscle fibers. J Gen Physiol 63, 509–530 (1974)
41. Matsubara I, Millman B M: X-ray diffraction studies on cardiac muscle. In: The Physiological Basis of Starling's Law of the Heart. Ciba Foundation London 1974, pp 31–41
41a Winegrad S: Resting Sarcomere Length-Tension – Relation in Living Frog Heart. J Gen Physiol 64, 343–355 (1974)
42. Pollack G H: Maximum velocity as an index of contractility in cardiac muscle. Circ Res 26, 111–127 (1970)
43. Brutsaert D L, Claes V A, Goethals M A: Effects of calcium on force-velocity-length relation of heart muscle of the cat. Circ Res 32, 385–392 (1973)
44. Henderson A H, Brutsaert D L: Force-velocity-length relationship in heart muscle: lack of time-independence during twitch contractions of frog ventricle strips with caffeine. Pflügers Arch 348, 59–64 (1974)
45. Brutsaert D L, Sonnenblick E H: Cardiac muscle mechanics in the evaluation of myocardial contractility and pump function: problems, concepts, and directions. Prog Cardiovasc Dis 16, 337–361 (1973)
46. Wise R M, Rondinone J F, Briggs F N: Effect of calcium on force-velocity characteristics of glycerinated skeletal muscle. Am J Physiol 221, 973–979 (1971)
47. Goethals M A, Adèle S M, Brutsaert D L: Contractility in mammalian heart muscle: calcium and osmolality. Circ Res 36, 27–33 (1975)
47a DeClerck N M, Claes V A, Brutsaert D L: Force-velocity relations of single cardiac muscle cells: calcium dependence. J Gen Physiol 69, 221–241 (1977)
47b Edman K A P, Kwong J C: The force-velocity relationship in vertebrate muscle fibres at varied tonicity of the extracellular medium. J Physiol 269, 255–272 (1977)
47c Lehman W: Thick-filament-linked calcium regulation in vertebrate striated muscle. Nature 274, 80–81 (1978)
47d Maughan D W, Low E S, Alpert N R: Isometric force development, isotonic shortening, and elasticity measurements from Ca^{2+} activated ventricular muscle of the guinea pig. J Gen Physiol 71, 431–451 (1978)
47e Schwartz K, Bouveret P, Sebag C, Leger J, Swynghedauw: Immuno-chemical evidence for the species-specificity of mammalian cardiac myosin and heavy-meromyosin. Biochem Biophys Acta 495, 24–36 (1977)
47f Jewell B R: A reexamination of the influence of muscle length on myocardial performance. Circ Res 40, 221–230 (1977)
48. Ross Jr J, Covell J W, Mahler F: Contractile responses of the left ventricle to acute and chronic stress. Eur J Cardiol 1, 325–332 (1974)
49. Abbott B C, Gordon D C: A commentary on muscle mechanics. Circ Res (in press)
50. Van den Bos G C, Elzinga G, Westerhof N, Noble M I M: Problems in the use of indices of myocardial contractility. Cardiovasc Res 7, 834–848 (1973)
51. Rushmer R F: Initial ventricular impulse – a potential key to cardiac evaluation. Circulation 29, 268–283 (1964)
52. Noble M I M: Problems in the definition of contractility in terms of myocardial mechanics. Eur J Cardiol 1, 209–216 (1973)
53. Paulus W J, Claes V A, Brutsaert D L: Physiological loading of cardiac muscle. Circ Res 39, 42–53 (1976)
54. Mirsky I: Left ventricular stresses in the intact human heart. Biophys J 9, 189–208 (1969)
55. Paulus W J, Claes V A, Brutsaert D L: Physiological loading of isolated feline cardiac muscle: The interaction between muscle contraction and vascular impedance in the production of pressure and flow waves. Circ Res 44, 491–497 (1979)

Dirk L. Brutsaert, M.D.,
Professor of Physiology,
Department of Physiology,
University of Antwerp,
Groenenborgerlaan, 171,
Antwerp (Belgium)

Adv. cardiovasc. Phys., vol. 5 (Part III), pp. 14–33 (Karger, Basel 1983)

Apexcardiology: Methodological Aspects and Its Role in the Assessment of Myocardial Contractility

J. L. Willems
University Hospital St. Rafael, Leuven, Belgium

Contents

Abstract

The important methodological aspects of the apexcardiography and its potential use in the assessment of myocardial "contractility", are highlighted. An overview of the information content and a discussion of the genesis of the diastolic and systolic deflections of the apexcardiogram are provided in the first section. The transducers and methods employed for the recording of sub-audible precordial vibrations are described in the second section. They can be subdivided into displacement and pressure measuring systems. The frequency responses and the calibration techniques are stressed. Computerised wave pattern recognition, and programs for the calculation of systolic time intervals and "non-invasive" measures of myocardial-contractility are presented in the third section. It is concluded that indices derived from (dp/dt)/P ratios of the apexcardiogram are promising for the non-invasive evaluation of left ventricular function.

I. Apexcardiography: Basic Contents and Correlations

A. Introduction

The first apexcardiogram or graphic registration of the pulsation of the heart against the thoracic wall was made in 1861 by Chauveau and Marey. They recorded this external indirect tracing simultaneously with pressure curves from different chambers of the heart in horses, dogs and other small animals and even in a few primates.

Using a capsule covered with a membrane and connected directly or indirectly by tubing to a lever, Marey (1863) also recorded on smoked paper precordial tracings of man. The Marey system or sphygmograph was subsequently modified by different authors and changes were made, mainly in the tambour and lever system.

Interest in the graphic registration of the apex beat has been widespread in the second half of the nineteenth century and the early 1900's, prior to the invention of electrocardiography. Variability in the records, however, was one of the main reasons why interest in apexcardiography progressively declined from 1920 on, despite the efforts by the school of Pachon. As early as 1902 Pachon stressed necessity of recording the apex impulse in the left recumbent position at the discrete place of maximal pulsations in order to obtain reproducible tracings.

The introduction of piezo-electric microphones by Miller and White (1941) and the use of electronic amplifiers simplified markedly the technique of apexcardiography and reinaugurated its acceptance in clinical medicine. Especially the work of Hartman and Snellen, from Leiden, The Netherlands, in the mid-fifties revived interest in this field.

B. Recording Procedures

For the recording of an apexcardiogram the subject, human or animal, is placed in the left lateral decubitus position and the point of maximal cardiac impulse, also called apex beat, is determined by palpation. In humans raising the left arm may help in locating the apex beat. This is most often in the fifth intercostal space, slightly to the left of the mid-clavicular line. The term "apex beat" has been avoided by some authors. It may indeed invoke the theory that the heart's impulse is always occasioned by the impact of the apex of the heart against the chest wall. This is probably incorrect, since especially in cases where the "apical impulse" is felt in the 4th intercostal space, it is certainly not the apex but merely the body of the left ventricle or septum that is felt beating under the fingers. The term apexcardiography however is generally accepted in the literature and will also be used throughout this chapter.

The apexcardiogram is conventionally recorded under oscilloscopic control in apnea at the end of expiration or in some patients at the beginning of inspiration, depending on the morphology of the curves. Indeed although the point of maximal impulse identifies the exact place for application of the pickup further minor adjustment of location and sometimes of the position of the patient may be necessary in order to obtain a tracing with a dominant positive systolic wave, a clear O point and clear initial systolic lift, as represented in Figs. 1 and 2.

The technique of recording the apexcardiogram (ACG) requires a certain experience, but is still relatively easy. One should be aware, nevertheless, that artifactual curves occur when the pickup is not placed directly over the center of the apex. Mirror image curves will be recorded in the zone of retraction around the apex and composite tracings farther in the periphery.

In the left lateral decubitus, however, reproducible curves can be recorded as stressed by Pachon (1902) and acknowledged by many investigators since. The left lateral position brings the heart closer to the chest and in that position the heart probably remains in contact with the thoracic wall throughout the whole cardiac cycle.

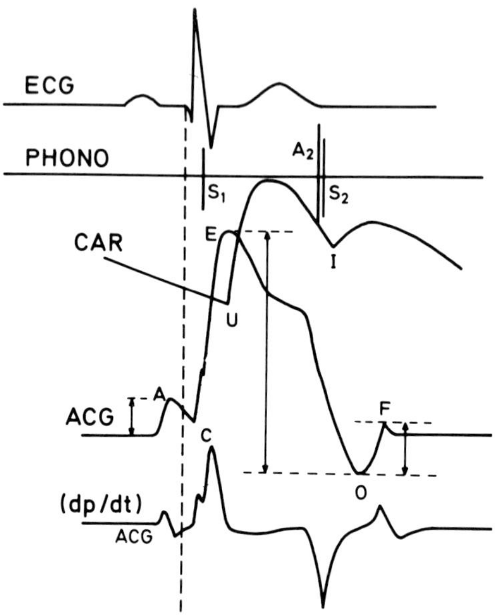

Fig. 1 Schematic representation of a lead of the electrocardiogram (ECG), the phonocardiogram (PHONO), the carotid tracing (CAR), the apexcardiogram (ACG) and its derivative (dp/dt). Abbreviations noted on this illustration are used throughout this chapter: Q for the onset of QRS; S1 and S2 for the first and second sound; A2 for aortic component of the second sound; U and I for the upstroke and incisura of the carotid pulse; A, C, E, O and F respectively for the A wave, onset, E and O points and F wave of the apexcardiogram. The vertical lines represent respectively the total height of the ACG and the amplitudes of the A and F waves. These waves are calculated as a percentage of the total systolic deflection.

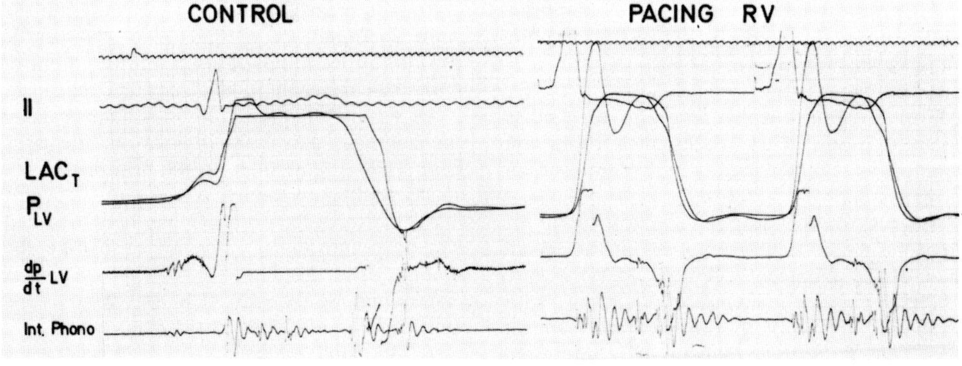

Fig. 2 Simultaneous recording of the left apexcardiogram (LAC), left ventricular pressure (P_{LV}) and its first derivative (dp/dt_{LV}), lead II of the electrocardiogram and the intracardiac phonocardiogram (Int. Phono) in a dog before (left) and during (right) pacing of the right ventricle. Paper speed = 250 mm/s. Note the simultaneousness of C and O points of LAC and P_{LV}.

Usually the apexcardiogram is recorded simultaneously with other indirect tracings, namely the electrocardiogram, phonocardiogram and central arterial pulse tracings. Temporal measurements provided by the apexcardiogram are indeed best appreciated in multichannel recordings, as demonstrated in Figs. 1, 2 and 3.

C. Description and Basic Contents of the ACG

1. Time Relationships

a. *The electromechanical interval*

Following the electrical depolarisation, mechanical activity starts in the heart after a variable interval. This period is referred to as the electromechanical lag or activation time. The onset of the systolic rise of the apexcardiogram, also called C point, denotes the onset of mechanical contraction of the heart. The C point occurs approximately 30 ms after the onset of the QRS complex of the ECG.

b. *The upstroke of the ACG*

After the C point the ACG presents a rapid systolic upstroke, reaching an end at point E at which the plateau of the curve is formed (Figs. 1 and 2). The initial upstroke is interrupted, in a high percentage of subjects, by small vibrations, a notch or a change in slope at the onset of the first heart sound. These changes point to the sudden changes in acceleration and vibration of the cardiohemic system occurring at the moments of closure of the atrioventricular valves.

c. *The E point*

The end of the straight systolic upstroke of the ACG has been called E point by Benchimol and Dimond (1963). When the rate of early ejection is unimpaired a rapid and sharp E point occurs close to the onset of systolic pressure rise in the aorta. Frequently, however, the E point is less well defined and the systolic upslope of the ACG continues to rise even after ejection of blood into the aorta has started.

Most normal subjects show a fall-off or plateau after the E point. When the ACG continues to rise after "point E" the morphology of the tracing has been labeled "sustained" (Sutton et al 1967, 1970).

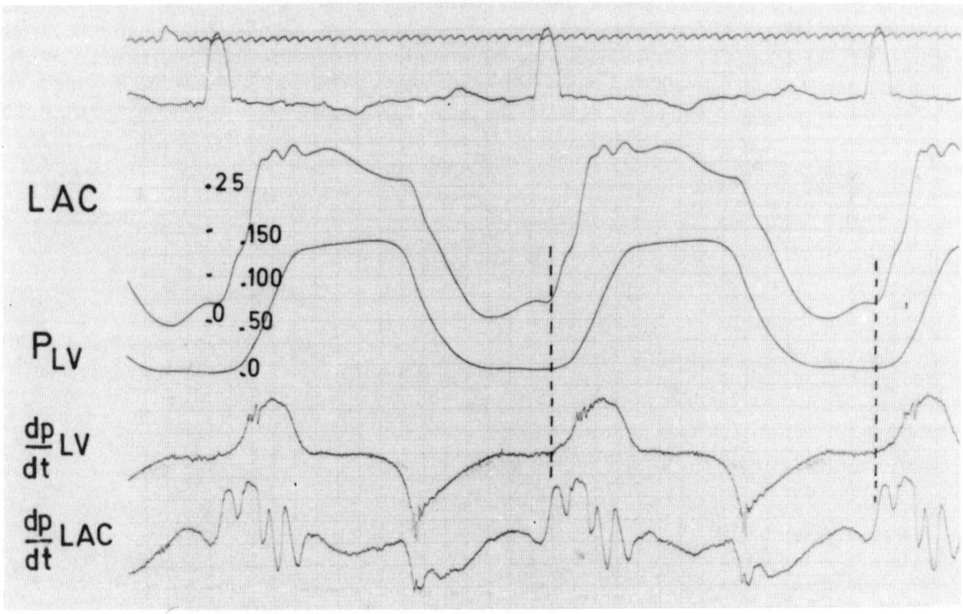

Fig. 3 Simultaneous recording of the left apexcardiogram (LAC), left ventricular pressure (P_{LV}) and their derivatives (dp/dt) in a dog. Note the presence of a definite deflection in the upstroke and also in the downslope of the ACG, which coincide repectively with the first and second sound. The initial upslope of the ACG occurs more rapidly than the pressure rise in the left ventricle.

d. *The isovolumic relaxation phase*

The decline of the systolic plateau starts about 30 to 40 ms before the beginning of the high frequency components of the second heart sound (A_2) and ends at the O point. The interval from A_2 till the protodiastolic nadir of the ACG or O point has been used as an estimate of the isovolumic relaxation time. This interval is delineated by the onset of the closure of the semilunar valves and the opening of the atrioventricular valves.

e. *The O point and the diastolic filling phases*

The O point of the ACG occurs close to the mitral valve opening. It also indicates the start of the rapid diastolic filling of the left ventricle. The F point (Fig. 1) marks the end of this rapid filling. The slow diastolic filling lasts till the onset of the atrial contraction, which is visible in the ACG by an A-wave. The atrial wave of the ACG occurs approximately 0.08 to 0.10 s after the beginning of the P wave of the electrocardiogram. The height of the A-wave is normally less than 15% of the amplitude of the total tracing.

f. *Temporal correlation of the ACG and LV pressure events*

The temporal relation between the internal left ventricular (LV) pressure and the apexcardiogram has been examined in man and in the dog (Willems 1973). One of the striking findings in the dog experiments was the marked synchronism of the upstroke and protodiastolic nadir of both the apexcardiogram and LV pressure curve, over a wide range of heart rates, throughout various acute hemodynamic changes or with different electrical conduction (Fig. 2). Also the peaks of the A and rapid filling waves coincided. The same could be demonstrated in man subjected to heart catheterization by means of high fidelity micromanometers. These results support the validity of the ACG for measurement of the electromechanical interval and as a reference curve in phonocardiography for timing intracardiac events.

2. Morphological Relationships and Correlations

a. *Diastolic waves*

One of the clinical useful signs of the ACG is the fact that it demonstrates abnormally large atrial components (A-waves) in cases with increased left-ventricular end-diastolic pressure, such as in patients with ischemic heart disease, aortic stenosis, myocardiopathy and other cardiac abnormalities. The ACG is particularly valuable in the assessment of left ventricular diastolic pressure events (Willems et al 1972). It has been demonstrated that changes in the O point level, total diastolic deflection and A-wave truly represent corresponding changes in LV pressure curves. When the myocardial and thoracic wall remain in close contiguity, which is the case in diastole,

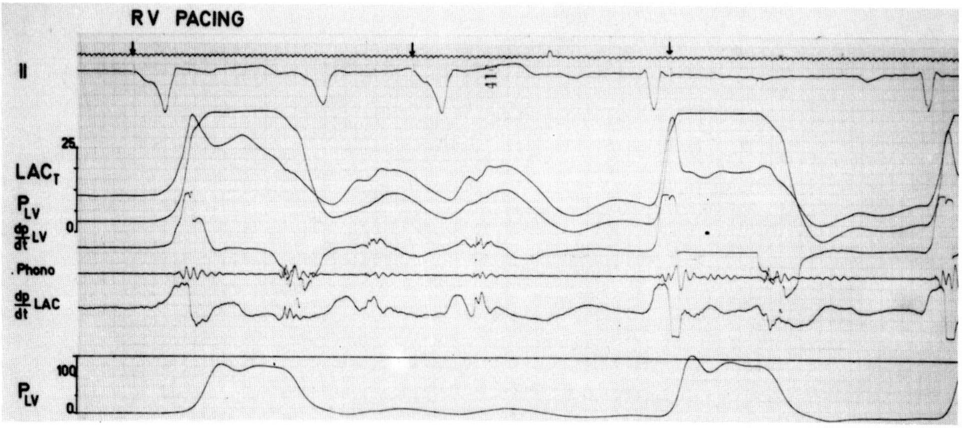

Fig. 4 Simultaneous recording of the left apexcardiogram (LAC), ventricular pressure (P_{LV}), their first derivatives, lead II of the electrocardiogram and the intracardiac phonocardiogram in a dog during electrical pacing of the right ventricle (RV). Note the simultaneity of C and O points of LAC and P_{LV} and the striking resemblance of morphology of the curves during extrasystoles during the 2nd and 3rd beat.

the intraventricular pressure propagates almost unattenuated through the chest wall. During that phase of the cardiac cycle other determinants influencing the apical impulse seem to be silent. This is not the case for the systolic component of the ACG.

b. *Systolic wave*

1. Total Height: Although a significant correlation ($r = 0.75$; $p < 0.001$) was found during dog experiments between percent changes in height of the systolic waves of the ACG and LV pressure (Fig. 4), during several acute hemodynamic interventions, it was clear that LV systolic pressure is not the only factor influencing the height of the ACG (Willems et al 1972). Variations in heart volume indeed markedly influenced ACG height, possibly by a better coupling of the heart with the chest wall or by changes in ventricular wall tension. Other factors as alterations in spatial position, the left to right torsion movements of the heart, changes in stroke volume and recoil forces are secondary determinants of the systolic deflection of the ACG. These factors are difficult to quantitate. They have so far only been qualitatively studied or analysed in a deductive way.

2. Systolic morphology during extrasystoles: During extrasystoles contraction of the heart can be assumed isovolumetric, when peak left ventricular pressure does not exceed aortic diastolic pressure. The myocardial wall probably remains in close contact with the chest during the whole cardiac cycle under these conditions. Pressure of the heart on the chest in such circumstances closely resembles left ventricular intracardiac pressure (Fig. 5). The same holds true for the initial, isovolumic phase of a normal contraction. During this phase, it has been shown that the external circumference of the heart slightly increases.

3. Systolic patterns in "cardiac" disease: Mainly Sutton and associates (1967, 1970) have studied the relationship of left ventricular hemodynamics with three types of systolic plateau's of the ACG, which they called the "normal", "hyperdynamic" and "sustained" patterns.

The "hyperdynamic" pattern resembles the normal outward movement but is exaggerated in height as demonstrated by quantitative apexcardiography. The hyperdynamic type is found in various conditions such as valvular mitral regurgitation and thyrotoxicosis, where increased volumes of blood are being handled by the left ventricle, although at normal pressure.

The sustained type occurred frequently in patients with low ejection fraction and an increase in end-diastolic volume, such as in cases with cardiomyopathy, long-standing val-

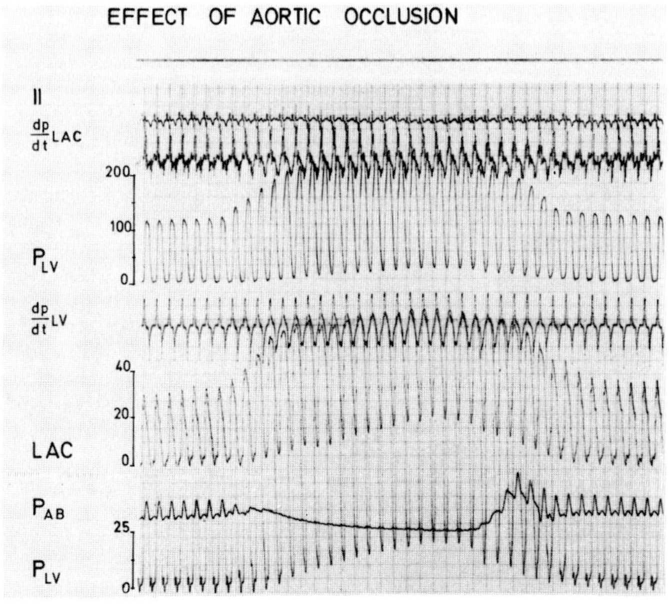

Fig. 5 Effect of ascending aortic balloon occlusion and release on the left apexcardiogram (LAC) recorded by means of a pressure measuring transducer and left ventricular pressure (P_{LV}) and their first derivatives (dp/dt). Soon after initiation of the occlusion, the pressure in the brachial artery (P_{AB}) drops, whereas LV systolic and early diastolic pressure increase. There is a progressive, almost equal rise in total systolic deflection and O point of the calibrated apexcardiogram. (scale = mm of Hg).

vular disease or myocardial infarction (Sutton et al 1970). In normal conditions a diminution in volume during systole causes a decrease in the force with which the heart presses upon the chest. In patients with impaired LV function changes in volume are markedly diminished. This is probably the reason why the morphology of the ACG more closely resembles LV pressure in such patients.

Higher up we already stated that left ventricular pressure per se is certainly not the only determinant of the systolic wave of the ACG. The genesis of the systolic component is much more complex. The heart may even in the left recumbent position cause traction upon the chest and result in a negative systolic deflection, at a time when LV pressure is rising. A systolic retraction in the ACG can be observed in idiopathic hypertrophic stenosis, in patients with systolic non-ejection clicks and late systolic murmurs and in cases with constrictive pericarditis.

c. *First time derivatives*

A significant correlation between simultaneously recorded first time derivatives of the ACG and LV pressure has been demonstrated by several authors (Reale 1967, Willems et al 1971, 1972, 1973, Denef et al 1974). The similarity in the rate of rise of both curves is especially striking during the isometric phase of contraction. These findings have prompted several investigators to compare in detail new indices derived from these first derivatives for the assessment of myocardial contractility, as will be described later in this chapter.

II. Methodology

A. Introduction

The frequency spectrum of vibrations transmitted by the heart and vibrating blood vessels to the precordium extends from 0 to over 1500 Hz, while the range of amplitudes is of the order of 100 to 120 db, decreasing approximately 6 to 15 db/octave with increasing frequencies (Berson and Pipberger 1966, Zalter et al 1959). All the techniques designed for the detection of precordial vibrations sacrifice a portion of the total information input in order to bring out the specific area of interest. These techniques restrict data output by their inherent qualities (frequency response, dynamic range) or by signal conditioners (filters, differentiators, integrators), which modify the data in a controlable manner. Phonocardiography gives a graphical display of audible precordial vibrations as revealed by auscultation. Apexcardiography on the other hand directs itself to the zone of subaudible frequencies, below 30 Hz, as detected by palpation and inspection of the anterior chest wall. Since the amplitudes of infrasonic vibrations are so much greater than those of sounds and murmurs, recording of subaudible frequencies needs no filtering for excluding heart sounds. In phonocardiography on the contrary, filters are needed to eliminate the dominating low frequency subaudible components.

Techniques for recording low frequency vibrations on the chest wall, comprising almost all the total vibratory energy, have received increasing attention in the past few decades.

B. Basic Recording Techniques

1. Recording of Precordial "Movement"

Subaudible vibrations have been recorded as displacement, velocity or acceleration of the thoracic wall.

a. *Displacement measuring methods*

The detection and measurement of chest wall displacement or movement has been performed by means of contacting and non-contacting transducers.

Experience is greatest with contacting transducers. The most commonly used method depends on air displacement in a closed system, namely a capsule of Marey or funnel connected by air tubing to a piezo-electric crystal microphone, as introduced by Miller and White (1941), or to a sensitive DC measuring pressure strain gauge (Kesteloot et al 1969). Among the types of transducers that need no air transmission are the condenser and carbon microphone (Brecht and Boucke 1952), the electromanometer (Johnston and Overy 1951), the accelerometer (Landes 1940, Mounsey 1957), the differential transformer (Schneider and Klunhaar 1961), the electromagnetic pickup (Rosa and Luisada 1959) and the variable resistance and inductive devices (Ueda et al 1962, Watson et al 1968).

To eliminate possible distorsion of displacement measurement due to contact with the chest wall, non-contacting transducers have been developed using body-pickup capacitance (Groom and Boone 1956, Podolak et al

1969), electromagnetic changes (Vas 1967) and fiberoptics (Cole and Conn 1971) or photoelectric detection of light reflected from the vibrating surface (Lohr et al 1963, Berry 1966). Berson and Pipberger (1966) reported on a photoelectric method that detects tridimensional motion of a lightweight target riding on the chest.

b. *Precordial velocity and acceleration tracings*

Velocity and acceleration of the thoracic wall have been studied in addition to true displacement, mainly in the 1950's when emphasis turned away from body- to precordial ballistocardiography (Agress and Fields 1959, Edson et al 1954, Hollis 1958, Rosa and Luisada 1959). It resulted in a variety of instruments, a multitude of wave forms and a variable nomenclature. Several investigators, studying precordial ballistocardiography and the related fields of accelerography and vibrocardiography have demonstrated similarities (Rosa and Luisada 1959) and even a basic mathematical relationship between these tracings (Agress and Fields 1959). Their clinical significance, however, has as yet to be properly delineated. The wave forms obtained by these systems depend mainly on their frequency characteristics, most showing a sharp drop in response below 30 Hz and almost no output below 4 Hz.

c. *Relative versus "absolute" displacement*

Conventional apexcardiography provides a record of the apex beat, measuring displacement in relation to the surrounding chest wall. It detects motion of the skin and soft intercostal tissues with respect to the rim of the sensing capsule or funnel. The capsule or sensing device is held by hand over the concise point of maximal pulsations with the patient in left recumbent position. In this respect, "conventional" apexcardiography records relative displacement that is the movement of a point in an intercostal space in relation to the surrounding area of the chest wall. It is theoretically possible that if every single point of the entire precordium moved throughout the cardiac cycle simultaneously and to the same extent, that no motion should be detected by apexcardiography (Bancroft and Eddleman 1967, Dressler 1937, Johnston and Overy 1951). Due to the large differences in stiffness of the ribs and intercostal tissues, this is only a theoretical hypothesis.

Dressler (1937) first suggested the possible importance of recording absolute precordial displacement. He arranged his instrument in a fixed support attached to the bed. This set-up has been used by Eddleman (1953) and by Mounsey (1957), in England, in propagating kinetocardiography and impulse cardiography respectively. Eddleman et al (1953) used a stiff metal bellow mounted on a crossbar, attached to the bed, and connected by air transmission to a piezo-electric transducer. Kinetocardiograms are records of precordial motion, in fixed locations corresponding to the positions of the precordial electrodes (V1 to V6) of the standard 12 lead ECG, with the patient being in the supine position.

Kinetocardiography has not gained much popularity since multiple tracings with varying wave forms and often redundant information have to be recorded. The clinical value of ACG recording for the temporal and mechanical assessment of left ventricular "events" however has been proven by many clinical studies and also by the simultaneous recording of ACG and intracardiac pressures (Willems et al 1971). An example of an apexcardiogram and its time derivative together with other curves are given in Figs. 1 and 2.

2. Recording of Precordial Pressure Variations

Attempts to record and quantitate precordial vibrations as pressure changes are scarce (Dunn and Rahm 1952, Gleichmann et al 1968, Willems et al 1971). As pointed out by Otto Frank (1910), there are basically two systems for graphic recording, namely displacement and force transducers, for isometric and isotonic measurement respectively. Recording of apexcardiograms with pressure (force or tension) transducers in order to measure real pressure changes has only recently been introduced (Gleichmann et al 1968, Willems et al 1971).

C. Frequency Spectrum of Low Frequency Precordial Vibrations and Its Implications on the Transducer Systems in Apexcardiography

1. "High" Frequencies

Repetitive non sinusoidal pulses, as precordial vibrations are, may be considered as a composite of various sinusoidal components. Mathematically they can be represented by Fourier analysis as the sum of a fundamental

Table 1 Amplitude and phase measurements of the fundamental frequency component and different harmonics obtained by Fourier analysis of 16 apexcardiograms, from 10 normal subjects ranging in age between 19 and 35 years. The mean heart rate was 67 beats/min (SD: 9.8; range 52–94). The apexcardiograms were recorded with a DC transducer linear to 100 Hz.

	Fundamental Frequency	Harmonics 1	2	3	4	5	6	7	8	9	10
Amplitude	100	41.2	17.4	19.4	9.8	5.8	3.4	3.0	1.6	1.5	0.8
(mean ±SD)		(12.5)	(7.8)	(6.9)	(7.5)	(1.6)	(1.5)	(2.2)	(1.3)	(1.2)	(0.6)
Phase	215	258	229	252	183	244	206	186	222	160	118
(mean ±SD)	(69)	(26)	(56)	(102)	(125)	(100)	(109)	(105)	(114)	(68)	(82)

sine wave and its harmonics. The possibilities and limitations of this method have been clearly discussed by Burger and Koopman (1949). Fourier of frequency analysis was used in our investigations to separate and to quantify each component of the apexcardiogram in 10 normal subjects. The apexcardiograms were recorded by means of a Marey capsule attached to a DC sensitive pressure wave transducer. The characteristics of this system have previously beein defined. Results of Fourier analysis of apexcardiograms in the 10 normal subjects are presented in Table 1. From this table it is apparent that the intensity of the 10th harmonic was on the average less than one percent of the amplitude of the fundamental frequency. The average frequency of the 10th harmonic of the apexcardiograms presently studied was 11 Hz. With a heart rate of 120 beats/min, the 10th harmonic has a frequency of 20 Hz. These findings are very important for the determination of the required frequency response of various transducers in apexcardiography. Actual frequency composition of some pulse wave tracings using Fourier transformation has previously been published in a few subjects by Landes (1942), Foulger et al (1947), Burger and Koopman (1949), Abbott (1969) and by Kesteloot, Willems and van Vollenhoven (1969). Above 15 Hz the amplitudes of frequencies diminish rapidly and become less than 1/100th of the amplitude of the fundamental frequency.

From the foregoing discussion it may be concluded that for reliable recording of low frequency vibrations the mechanocardiographic transducers should have a flat amplitude frequency response up to 30 Hz. This is the 10th harmonic of a heart beating 180 times per minute. However systems linear to 50 and even 100 Hz are desirable for clinical work, since in some patients with heart disease higher frequencies may be more intense than in normal subjects.

2. "Low" Frequencies and Time Constant of the Recording System

As for the lower end of the frequency spectrum, ideal apexcardiographic recording systems should measure vibrations from 0 Hertz on, which implies DC recording (Bancroft and Eddleman 1967, Johnson et al 1971, Kesteloot et al 1969). In the era preceding 1941, the time when piezo-electric microphones were introduced, practically all transducers used in apexcardiography indeed met this requirement. A wide variety of systems is nowadays used to record the apexcardiogram, most of them having a finite instead of an infinite time constant. The RC time constant of a transducer can be determined by measuring the time required for a 63.2% decay of the transducer output while it is subjected to a step function. Although Miller and White (1941) had warned against the fallacies of too short a time constant of piezo-electric microphones, many users failed to recognize the fundamental artifactual changes that tracings, recorded with systems with inadequate time constant, present. Numerous articles have been published mainly in the American literature with apexcardiograms distorted and differentiated by too short a time constant apparatus. This restricts comparison of data from different laboratories not only concerning the morphological appearance of the curves but also concerning timing of different waves and intervals (Willems et al 1971).

The fallacies of devices with too short a time constant have been stressed recently by many

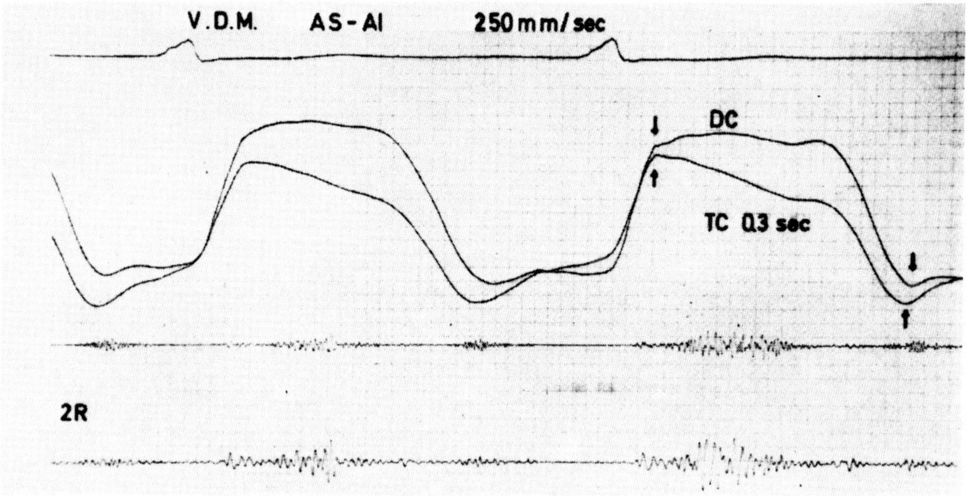

Fig. 6 The O point of the LAC recorded with a TC of 0.3 s precedes the same point on the DC-curve by 12 ms. This time-shift varies from beat to beat due to slight changes in the form of the curve. High frequency components of both curves, as the summit of the rapid filling wave and the end of the upstroke occur simultaneously. Important changes in slopes are also introduced by the TC. (Fig. 2 of paper by Kesteloot, H., Willems, J., van Vollenhoven, E., published in Acta Cardiologica 24, 150 (1969). Permission for reproduction was granted by the publisher and the authors.)

authors. The shorter the time constant of a system the more rapid the base line will be reached, even when a constant displacement of pressure is maintained. One will never find real plateaus in ACG tracings recorded with a time constant of 0.3 s, a time constant formerly used by many investigators in the U.S.A. (Fig. 6).

These tracings always present systolic waves with a steep downslope, exaggerated rapid filling waves and time shifts of the O point. Due to differentiation the O point of the ACG, recorded with devices with too short a time constant, will occur earlier. No phase shifts will be introduced through DC recording apparatus. It has been demonstrated however that in common clinical use, tracings recorded with a time constant of 3 s are not different from DC tracings (Kesteloot et al 1969, Willems et al 1968).

Differences between DC curves and curves obtained after differentation with a TC of 1 s are minor when the heart rate is above 70/min. Below this heart rate, differences in diastolic plateaus will be visible even in curves recorded with a TC of 1 s which will present some artifactual sloping up feature.

3. Practical Frequency Characteristic Requirements

Johnson et al (1971) pointed to the fact that components of the recording system, other than the transducer, may limit the degree of fidelity that can be achieved. Firm binding or application of the transducers extended the upper frequency range of their system. They further warned to the fact that the amplifiers of the recorder itself may distort waveforms. Leaks may also limit frequency response to a marked extent (Kastor et al 1970). Through an air leak pressure in the recording system escapes and spontaneous decay of the curve occurs similarly as through the use of AC filters, causing significant amplitude attenuation in the low frequency range, especially below 1 Hz.

For apexcardiography an overall time constant of at least 2.5 s should be recommended for the total recording system. A system with a time constant of 2.5 s and a 6 d*b*/octave low-frequency roll-off is linear down to 0.064 Hz and is a reasonable compromise between accurate waveform representation and suppression of low-frequency base-line shifts (Johnson et al 1971).

D. Quantitation in Apexcardiography

1. Calibration on Precordial Displacement Tracings

Calibration of actual precordial movement as a means of non-invasive assessment of left ventricular (LV) hemodynamics has been propagated by Sutton and Craige (1967, 1970). The utilities of precordial movement in providing information concerning LV hemodynamics has been discussed in paragraph 1. Sporadic information on calibrated displacement tracings has also been published by Landes (1940), Eddleman et al (1953), Gillam et al (1964), Berson and Pipberger (1966), Podolak et al (1969) and by Ikegaya et al (1971). Some investigators expressed the height of the ACG deflections in terms of a constant signal, corresponding to the movement of a calibrator (Sutton et al 1967); others have tried to provide absolute figures of displacement (Ikegaya et al 1971, Landes 1940, Landes 1942, Podolak et al 1969, Willems 1973). Absolute deflections of the ACG in 239 normals varied between 20 and 850 microns (Willems 1973). The 96 percentile upper limit equaled 650 micron. In patients with "hyperdynamic" apical impulses such as in hyperthyroidism and mitral incompetence, exaggerated deflections have been recorded (Craige 1971, Sutton and Craige 1967, Sutton et al 1970).

Quantitation of precordial displacement is influenced by many variables. Most of the variation of displacement records in normals can be attributed to the anatomic aspects of the chest wall or to the technical problems of recording and calibration. This is the main reason why quantitation of the apexcardiogram has not been widely accepted in clinical cardiology.

2. Calibration of Precordial Pressure Variations

Recording of the ACG with a pressure measurement system has recently been introduced (Willems et al 1971). The total deflection of the ACG, defined as the pressure difference in millimeters of mercury between the nadir of the tracing and the systolic peak, averaged 63.8 mmHg (*SD* 16.9) in 51 normal male subjects (Willems 1973). The coefficient of variation of 25.4% was less than the ones observed with the displacement recording systems, where values between 30 and 70% have been reported.

The low frequency precordial pressure fluctuations are much higher in intensity than the pressure variations of heart sounds. Values between 20 and 130 mmHg have been recorded, which corresponds to energy levels of 25,000 to 175,000 dynes per square centimeter (Willems 1973). Investigators using calibrated phonocardiography on the other hand, have demonstrated the energy of sounds between 10 and 100 cycles/s to vary between 2 and 200 dynes/cm^2 (Dunn and Rahm 1952).

The variables influencing the vibration energy at the precordium have been discussed in detail by some authors (Willems 1973). They are the vibrational output of the generator, the transmission characteristics and mechanical impedance of the chest wall, the mechanical coupling of the heart to the thoracic wall and the coupling of the transducer to the chest. Only the latter two variables can be influenced by the recording technique and pulse wave transducer.

As could be expected from clinical experience, patients with hypertension, valvular aortic stenosis and mitral insufficiency showed exaggerated apical impulses. The average height of the ACG pressure pulse in these patients varied between 87 (± 21) and 107 (± 26) mmHg which is statistically higher than in normal subjects (Willems et al 1971). Calibration of the ACG intends to quantitate the subjective feelings of the physician, palpating the precordium.

III. Apexcardiography and the Assessment of Myocardial "Contractility"

A. Introduction

One of the most important recent developments in the study of the mechanics of the heart has been the consideration of the heart as a muscle rather than as a pump. Over the past decade considerable effort has been devoted toward development of indices of myocardial contractility. Methods determining force-velocity-length characteristics of isolated skeletal and cardiac muscle (Brutsaert and Sonnenblick 1969) have recently been applied to the evaluation of the intact heart.

V_{max} indices have been calculated from simultaneous left ventricular pressure and volume determinations, using Laplace's law in the computing of wall tension and force. Volumetric data are cumbersome to obtain and are subject to a variety of errors, because the geometry of the ventricle is complex and difficult to assess accurately (Falsetti et al 1971).

Later it was found that indices derived from left ventricular pressure data alone correlated very well with the more elaborate ones involving angiocardiography. Correlations of these results with clinical history and diagnosis have shown that indices of contractility based on pressure data alone provide a sensitive indicator of myocardial performance (Falsetti et al 1971, Hugenholtz et al 1970). Such a procedure is extremely useful because it permits an assessment of contractility without the use of angiographic techniques. Although these methods provide much simplification, the need for strict non-invasive techniques for cardiac evaluation remains.

The advantage of non-invasive methods is that they can be used without any harm in patients too ill to be subjected to left heart catheterization and that they can be applied repeatedly in follow-up studies.

Measurement of the pre-ejection period (PEP), left ventricular ejection time (LVET) and LVET/PEP ratio has gained wide-spread attention in recent years under impulse of Weissler and associates (Weissler et al 1968). Other subdivisions of systole have been used for indirect evaluation of cardiac contractility. However most of these intervals are influenced by multiple variables, such as ventricular end-diastolic volume, peripheral resistance and the level of neurohumoral cardiostimulatory activity. Only with the exquisite control of these variables can the response in systolic time intervals be employed as a quantitative measure of myocardial contractility.

The time from the onset of ventricular contraction to the peak of the first derivative of the apexcardiogram has been advocated by Reale (1967), as a more direct simple method of detecting changes in myocardial contractile state. Since the introduction of calibrated apexcardiography, other indices, related to the magnitude of the derivative of the ACG, have been proposed (Gleichmann et al 1968, Willems et al 1972).

The ratio $(dp/dt)/P$ derived from ACG has recently been computed from the ACG in analogy to the ratio $(dp/dt)/P$ derived from ventricular pressure data (Denef et al 1973, Kesteloot et al 1971, Mirsky et al 1972, Motomura et al 1973, Pasternac et al 1971, Willems 1973). Measurement of the various parameters were obtained either by manual means or by computer pattern recognition and measurement programs with clearly defined criteria, and more recently by analog circuitry (Grossman et al 1971, Winter and Fabian 1973).

The non-invasive approach to "myocardial contractility" has been mostly pragmatic. Various indices derived from LV pressure data and reported by several authors as "measures of myocardial contractility" have been compared with similar indices derived from the ACG. These indices are related to, but certainly provide no unique measurement of myocardial contractility.

B. Methods

Myocardial contractility indices were derived from simultaneously recorded apexcardiograms and high fidelity left ventricular pressure curves in dogs, before and after administration of several vasoactive drugs, including calciumchloride, nembutal, isoproterenol and norepinephrine (Willems 1973). Similar studies were performed in humans during routine heart catheterization and in control subjects at rest (Willems 1973, 1974, Denef et al 1973, 1974).

1. Computer Methods and Programs for Wave Recognition and Measurement

In the dog experiments curves were recorded both on paper and on analog magnetic tape (Willems 1973, 1974). Five channels of information were recorded simultaneously: the electrocardiogram (lead II or lead I), the internal left ventricular phonocardiogram, the aortic and left ventricular pressures both with Telco micromanometers and the apexcardiogram, using a fluid-filled pressure measuring transducer.

Conversion of the analog data to digital form was carried out at a sampling rate of 250 samples/s over a period of 5.16 s using an IBM 1800 or HP 2100 computer. The peak-to-peak input voltage into the A to D converter of each channel was between 0.6 and 2 V (mean = 1 Volt).

Computer analysis of the ECG, phonocardiogram and carotid tracings aiming at recognition of the onset of QRS, the onset of the second heart sound and of the upstroke and incisura of the carotid pulse was performed using automatic wave recognition and measurements programs, details of which have been published (Kyle and Freis 1970, Willems 1974). The original computer programs have been implemented further to locate the C and O points of the ACG (see Fig. 1) and the onset of the first heart sound (Willems, Kyle and Freis, non published results). Algorithms for recognition of C and O points on the left ventricular pressure pulse and for determination of the height of the A wave were added (Willems 1973).

The C point was searched for in an interval starting 8 ms after the onset of the ECG and ending at 64 ms after Q. In case of conduction disturbances, especially left bundle branch block, both boundaries were increased with a variable amount depending on the manually determined QRS duration. A 9 point weighted smoothing was performed first. Next a search was made for four successive first differences greater than 6 mV for the ACG and greater than 4 mV for the LV pressure curve. This amounts to a slope limit of approximately 100 mmHg/s for the determination of the onset of the LV pressure curve, when the 100 mmHg calibration signal equals 1 V. If the search for a C point slope was positive, the first point by definition was taken as point C.

In case the above described criterion was not fulfilled within the specified range, the statement "cannot find slope change for C" was printed. In several hundred beats checks were made on manually and computer derived C point locations using large computer plots. Manually and computer derived C point coincided in almost all cases within a 3 sampling point or 12 ms interval. In the majority of comparisons the difference did not exceed plus or minus 8 ms.

The O point was computed as the lowest point of either the ACG or the LV pressure curve in a search interval ranging from the location of the peak of the R wave of the corresponding cycle till 100 ms before the next R wave peak of the ECG.

For the $(dp/dt)/P$ index, the first derivative was not recorded on a separate tape channel but was digitally derived using a first difference method.

The first derivative of the left ventricular pressure and of the apexcardiogram, at 4 ms intervals, was calculated by the computer as the difference of pressures measured 4 ms before and 4 ms after the point in question, divided by 8 ms. Thus at any point on the pressure time curve the first derivative may be represented by the formula

$$(dp/dt)_n = (P_{n+1} - P_{n-1})/(t_{n+1} - t_{n-1})/0.008$$

where P_{n-1}, P_n and P_{n+1} are any 3 consecutive pressures measured 4 ms apart at times t_{n-1}, t_n and t_{n+1}. Identical procedures were used by Mirsky et al (1972) who measured pressures at 5 ms and later at 3 ms intervals.

The ratio $(dp/dt)/P$ of the ACG was computed at every fourth ms over an interval of 100 ms starting at 12 ms after point C of the ACG. Negative ratios or ratios below 0.3 were not used in any computation (Willems 1973).

2. "Pragmatic" Indices of Myocardial "Contractility"

The following "myocardial contractility" indices derived from apexcardiograms and left ventricular pressure tracings were investigated (Willems 1973).

1. Maximum dp/dt, expressed in mmHg/s
2. The time from C to peak dp/dt, in ms
3. Maximal or peak $(dp/dt)/k.P$ ratio also called peak VCE, expressed in muscle lengths/s, where P equals "instantaneous" and "total" pressure in case of LV pressure data and only "total" pressure in case of apexcardiographic tracings. For definitions see also Fig. 7.
4. V_{max} indices, computed by streight line extrapolation, using "total" and "instantaneous" pressure and also by polynomial fitting on developed pressures. (see Fig. 7 for further explanation).

The derivation of $(dp/dt)/(kP + C)$, $(dp/dt)/kP$ and $(dp/dt)/P$, as approximations of contractile element shortening velocity during isovolumic systole has been discussed by several authors (Falsetti et al 1971; Hugenholtz et al 1970).

It should be reminded however that all these measures provide only partial approximations of the true maximal velocity of contractile fiber shortening. The indices derived from the ACG are a further extension of the pragmatic approach in the assessment of myocardial contractility and are not based upon clearly defined theoretical considerations (Willems 1973). An elaborate formulation of the "force-velocity of contraction" concept based on 3 dimensional muscle-element has recently been given by Ghista et al (1973). Application of the

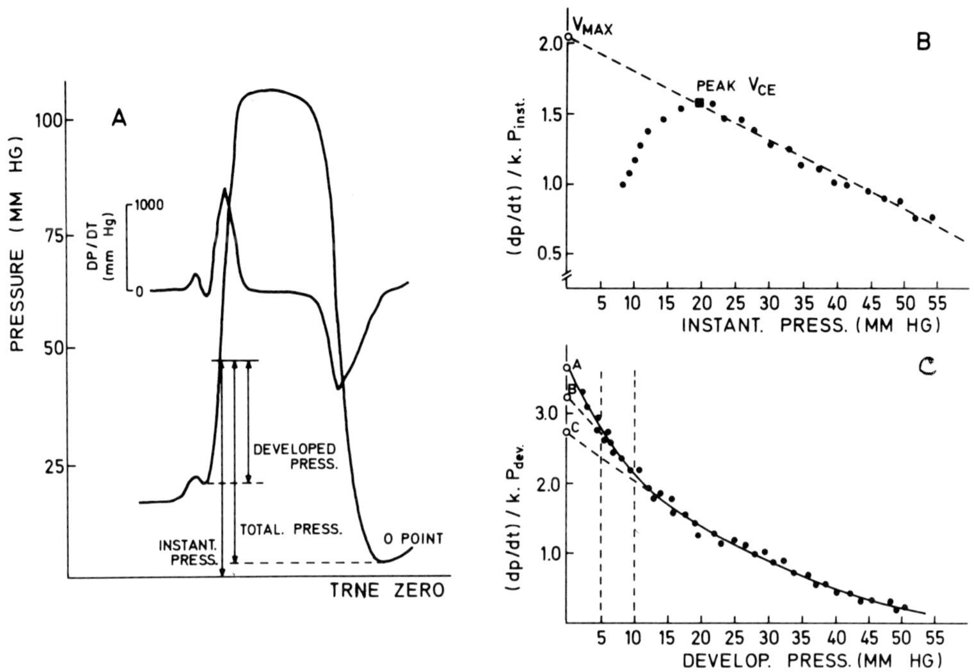

Fig. 7 A. Reference levels for determination of instantaneous, total and developed pressure are represented. They are the zero level, which refers to an external point 5 cm below the angle of Louis (manubriosternal articulation), and respectively the O and C points of the left ventricular pressure curve. In apexcardiography no true zero reference level exists. Consequently only "total" and developed pressure can be measured on the ACG.
B. Pressure-velocity curve relating instantaneous pressure (abscissa) and the ratio $(dp/dt)/k.P_{inst}$, also called V_{CE} or "velocity of contractile elements" (ordinate). Peak V_{CE} or maximal $(dp/dt)/k.P_{inst}$ is denoted by a square. V_{max} index is obtained by linear fitting of data points of the descending part of the curve. Identical procedures are followed for total pressure.
C. Pressure-velocity curve based on developed pressure. Data points of several beats were pooled and are depicted on this graph. V_{max} is obtained by polynomial fitting of the second degree and is represented by the intercept with the ordinate, being zero developed pressure. Three different V_{max} indices have been calculated on data from such a composite of beats: A, obtained by fitting of all data points; B and C, obtained by using only data above respectively 5 and 10 mmHg.

principles of such a model to the intact heart will be discussed elsewhere in this book.

V_{max} and V_{CE} data are expressed in units per s, denoted as s^{-1}. Since P is present in both the nominator and denominator, sensitivity at which ACG curves are recorded does not influence the result. V_{max} and V_{CE} values therefore can also be digitally calculated on non-calibrated ACG recordings (Willems 1973).

3. Use of Analog Circuitry

Electronic differentiation of the logarithm of intraventricular pressure provides a continuous on-line record of $(dp/dt)/P$. Such a method utilizing simple logarithmic amplifier circuitry has recently been applied by several authors on left ventricular pressure data (Grossman et al 1971, Winter and Fabian 1973) and on apexcardiographic tracings (Denef et al 1975).

The basis of the above circuitry is as follows. A rule of differential calculus is $(d/dy)\ln x = (dx/dy)x^{-1}$ Accordingly, $(d/dt)\ln P = (dp/dt)p^{-1}$. The differentiated output of the log amplifier circuit therefore provides a continuous readout of $(dp/dt)P^{-1}$.

The output of the log amplifier, i.e. the $\ln$ of LV pressure is differentiated electronically by an RC circuit with a time constant of 0.5 ms, in order to achieve this goal. In the case of the ACG, one needs calibration of the original signal and of its derivative and an electronic follower in order to determine the level of the O point (Denef et al 1975). With analog circuitry one can easily determine peak VCE, but for the other indices further calculations have to be performed as demonstrated in Fig. 7.

C. Results

1. Comparison of Computer and Manually Derived Time Intervals and "Myocardial Contractility" Indices

Correlation between manually and computer derived LVET, Q-A2 S1-S2, C-O and Q-upstroke intervals from 51 normal subjects resulted in highly significant correlation coefficients, as demonstrated in Table 2. Paired t-tests based on individual observations obtained by both methods showed no significant differences, except for the C-O interval. The C-O interval obtained by the computer wave recognition and measurements programs was on the average 9 ms longer.

Correlations of shorter intervals and of ratios showed smaller, but still statistically significant correlations. As a result of the differences in PEP and C-O intervals, also ratios derived from these measurements were slightly different (Table 2). The mean of all individual differences between computer and visually derived values was close to zero for most measurements. Averages and standard deviations of absolute differences are given in the last two columns of Table 2. They summarize how close computer and manually determined results agreed. Regression analysis on computer and manually derived peak V_{CE} values resulted in a correlation coefficient of 0.97 ($P < 0.001$) and a standard error of estimate of 0.05 units. The correlation coefficient equaled 0.96 for dp/dt magnitude, 0.89 for LV systolic pressure and 0.97 for heart rate.

2. Comparison of Non-Invasive with Invasive Indices of "Myocardial Contractility"

Myocardial contractility indices derived from simultaneously recorded apexcardiograms and high fidelity left ventricular pressure curves in dogs, before and after administration of calcium chloride, nembutal, isoproterenol and norepinephrine are presented in Table 3.

Percent changes of V_{max} indices and peak V_{CE} from ACG and LV pressure data after various drugs occurred in the same direction and were of similar magnitude. The same was true for the other indices (Fig. 8).

Table 2 Comparison of paired observation derived by computer and manual methods

	Mean results			Differences in paired data				
Measurements	Computer (ms)	Manual (ms)	Corr. coeff. (r)	Mean diff.	S.D.	t test	Abs.mean diff.	S.D.
LVET	280 (23)	281 (22)	0.96	−0.84	6.6	−0.9	5.3	4.0
Q-A2	381 (32)	379 (29)	0.92	2.39	12.8	1.3	10.5	7.6
S1-S2	314 (29)	312 (25)	0.86	2.19	14.7	1.1	12.0	8.7
C-O	453 (41)	444 (37)	0.85	8.80	21.4	2.9	18.1	14.2
Q-upstroke	136 (17)	135 (15)	0.82	1.16	9.7	0.8	7.6	6.0
PEP	102 (17)	98 (16)	0.70	4.00	12.7	2.1	10.4	8.2
PEP/LVET	0.365 (0.063)	0.349 (0.058)	0.66	0.016	0.050	2.1	0.040	0.033

Data were obtained from 50 subjects with a mean heart rate of 75.9 (S.D. 14.5) beats/min. All intervals are given in ms.

Table 3 Mean and standard deviation of representative values of "myocardial contractility" indices derived from simultaneously recorded apexcardiograms and left ventricular pressure curves in dogs.

		V_{max} indices				Peak V_{CE}	
				Developed P			
		Instant.P	Total P	Lin.	Poly 2nd	Instant.P	Total P
1. Control recordings ($n = 193$)	ACG		2.08 (0.50)	4.39 (0.77)	5.67 (1.16)		47.4 (7.8)
	LVP	1.75 (0.30)	1.91 (0.26)	3.34 (0.52)	4.23 (0.75)	1.64 (0.35)	60.5 (8.7)
2. Calciumchloride ($n = 94$)	ACG		2.46 (0.44)	4.93 (0.88)	6.09 (1.20)		54.2 (10.7)
	LVP	2.32 (0.51)	2.39 (0.44)	3.88 (0.58)	4.68 (0.77)	2.06 (0.57)	67.4 (12.5)
3. Nembutal ($n = 27$)	ACG		1.67 (0.27)	3.72 (0.71)	4.59 (1.20)		37.1 (4.4)
	LVP	1.51 (0.21)	1.66 (0.21)	3.27 (0.43)	4.39 (0.82)	1.40 (0.20)	53.3 (6.7)
4. Isoproterenol ($n = 36$)	ACG		2.61 (0.50)	4.54 (0.77)	6.26 (1.14)		59.0 (14.2)
	LVP	2.51 (0.42)	2.71 (0.43)	4.03 (0.39)	4.64 (0.65)	2.21 (0.59)	78.9 (18.7)
5. Volume load + norepinephrine ($n = 4$)	ACG		2.86 (0.24)	5.37 (0.36)	6.91 (0.81)		64.3 (11.5)
	LVP	2.35 (0.05)	2.77 (0.06)	4.23 (0.32)	4.90 (0.51)	1.71 (0.08)	75.3 (1.7)
6. Total group ($n = 382$)	ACG		2.25 (0.52)	4.47 (0.88)	5.79 (1.30)		49.7 (11.3)
	LVP	1.96 (0.54)	2.12 (0.50)	3.58 (0.62)	4.44 (0.75)	1.75 (0.50)	63.6 (13.5)

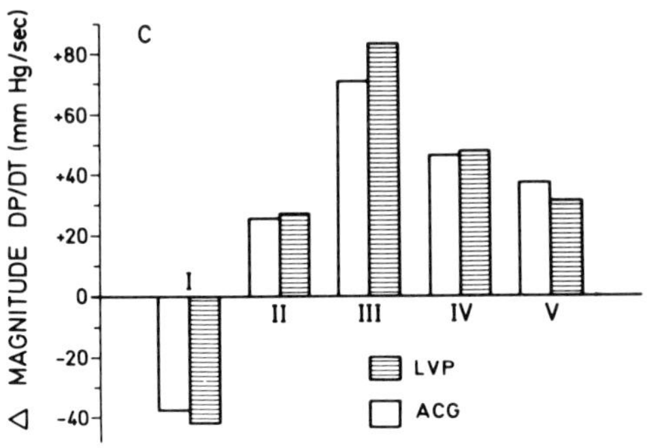

Fig. 8 Percent changes in magnitude of dp/dt from apexcardiographic and left ventricular pressure data in dogs. The roman numerals refer to different interventions one negative inotropic namely nembutal (I), and the others positive inotropic, norepinephrine (II), isoproterenol (IV) and calciumchloride (V).

Correlation between dp/dt magnitudes of calibrated ACG and LV pressure curves resulted in the highest correlation coefficient ($r = 0.76$; $n = 382$; $P < 0.001$), followed by peak V_{CE} ($r = 0.65$; $n = 382$) and V_{max} using total pressure ($r = 0.58$; $n = 382$).

Correlation of mean values obtained from the records of the different interventions ($n = 8$) gave very high correlation coefficient (r between 0.78 and 0.99). More detailed results have been published (Willems 1973).

In another series of experiments a correlation coefficient of 0.67 was obtained between peak V_{CE} derived from the calibrated ACG using total pressure and peak V_{CE} from simultaneously recorded LV data (Denef et al 1973).

3. Non-Invasive Measures of "Myocardial Contractility" in Normal Subjects

Mean values, standard deviations and coefficients of variation of peak V_{CE}, V_{max} indices using total and developed ACG pressure

Table 4 Normal values of "myocardial contractility" indices derived from the apexcardiogram in a group of 51 subjects.

	Mean	Standard deviation	Coeff. of variation	Normal lower (96 percentile) limit
V_{max} using total P (s^{-1})	1.80	0.42	23.3	1.20
V_{max} using developed P				
Linear extrapolation	2.83	0.59	20.8	1.84
Polynomial fitting	3.41	0.84	24.7	2.20
Peak V_{CE} (s^{-1})				
Formula $(dp/dt)/P$	39.6	6.9	17.0	29.0
Formula $(dp/dt)\ k.P$	1.24	0.21	17.0	0.90
Peak time (ms)	49.8	14.7	29.6	80 (65)*
Heart rate (beats/min)	77.0	14.7	19.1	—
A wave amplitude (percent)	7.2	2.4	33.3	—
Total diastolic deflection (percent)	17.3	5.0	28.9	—

*96 and 90 percentile upper limit.

variations and time to peak dp/dt of the ACG obtained by computer methods in 51 normal male subjects are presented in Table 4. They can be used as standards to which results in patients can be compared.

4. Preliminary Results in Patients

The normalized first derivative of calibrated displacement ($dD/dt/D_t$) apexcardiograms was definitely lower ($P < 0.005$) in patients with congestive cardiomyopathy (mean: 26.5; SD: 6.3; $n = 14$) than in normal subjects of similar age (mean: 33.8; SD: 7.0; $n = 30$) (Denef et al 1975). Lower mean values were also found in the same study in patients with ischemic heart disease (mean: 30.1; SD: 11.2). A statistically significant correlation existed between $dD/dt/D_t$ and left ventricular end-diastolic pressure ($r = -0.54$; $n = 42$) and also with the left ventricular ejection fraction (Denef et al 1975). All patients, except one, with values of $(dD/dt/D_t)$ max $\geq 34\ s^{-1}$ had normal ejection fractions (i.e. ≥ 0.64). Whereas all but one patient with a $(dD/dt/D_t)$ max $\leq 23\ s^{-1}$ (i.e. normal mean value minus 2 SD) had an ejection fraction below the lower limit of normal (< 0.64) (Denef et al 1975).

D. Discussion

The two basic objectives of this subchapter were the presentation (1°) of automatic analysis of the apexcardiogram, (2°) of non-invasive measures of "myocardial contractility".

Both method and measurement have only recently been introduced in apexcardiography and may be of great potential use in the future.

1. Automatic Analysis of the ACG

It has been demonstrated that computer wave recognition and measurements programs give sufficiently close results to manually derived measurement of systolic time intervals and various indices of myocardial contractility, to warrant their use in clinical practice. Average absolute differences of about 5 ms were present between computer and visually obtained intervals (Table 2), which amounts to about 2 to 3% of their total value. Such deviations are determined by the resolution capability of both computer and manual methods. Since the sampling rate, for technical reasons, was 250 samples/s resolution of the present computer method is 4 ms. The human eye has a similar resolution capability when tracings are recorded at 100 mm/s. In clinical phonomechanocardiography, however, most often a paper speed of 50 mm/s is used. Computer methods therefore may provide data of superior quality.

The advantage of the computer is further the speed with which it can quantitatively analyze records and make computations on a large number of beats. Manual determination of "myocardial contractility" indices and systolic time intervals are so tedious and time consuming that computer methods provide the only workable solution to these measurements.

2. Non-Invasive Measures of Myocardial Contractility

Various systolic time intervals and measurements related to the rate of rise of the apexcardiogram have been used to assess changes in rate of rise of LV pressure, in a non-invasive approach of studying "myocardial contractility".

Whereas systolic time intervals provide indirect evidence of changes in rate at which intraventricular pressure rises, more direct information in this respect can be gained from the first derivative of the apexcardiogram. A significant correlation between simultaneously recorded first time derivatives of ACG and LV pressure has been demonstrated by several authors (Reale 1967, Vetter et al 1972, Willems et al 1972, Willems 1973). The time interval from the onset of ventricular contraction to the peak of the first derivative of the ACG decreased significantly after isoproterenol, in normal subjects and in dogs, as a result of the increased contractile state, confirming observations of others (Fincker et al 1971, Reale 1967).

Calibration of the apexcardiogram using devices recording precordial movement and precordial pressure variations made it possible to include the magnitude of the first derivative in the evaluation of the contractile state of the heart. Positive inotropic interventions in dogs provoked a marked increase in peak dp/dt of both ACG and LV pressure, while negative inotropic interventions, vagal stimulation and propranolol had the opposite effect (Willems et al 1972). However, although standardization can reduce much of the intra-individual variability, total height of the curves and hence also the derivative, may vary according to the respiration phase and the position of the patient.

Use of dp/dt max of left ventricular pressure as an approximation of myocardial contractility is now being replaced by the normalized ratio $(dp/dt)/P$. It has been shown that the ratio of the rate of rise to the simultaneous ventricular pressure approximates, under certain assumptions, the contractile element velocity during isovolumic systole. The maximal no-load velocity of myocardial contractile elements (V_{max}) obtained by extrapolation of pressure velocity data, as well as peak V_{CE} were found useful in separating patients with cardiac failure from normals. In view of the high correlation between first derivatives derived from the ACG and LV pressure curves, it had been suggested that a similar ratio $(dp/dt)/P$ based on the ACG seemed worthwhile to investigate (Willems et al 1971). A significant correlation ($r = 0.67$; $P < 0.001$) could be demonstrated in dogs subsequently subjected to various hemodynamic interventions between the maximal ratio dp/dt divided by instantaneous P, from simultaneously recorded ACG and LV pressure (Kesteloot et al 1971, Motomura et al 1973, Willems 1973).

Recently Mirsky et al (1972) also demonstrated in man that the apexcardiographic "myocardial contractile" index $(t(dl/dt)/l)$ max. correlated well ($r = 0.72$) with peak V_{CE} determined from ventricular pressure curves in 18 subjects with various heart diseases.

In further studies extensive comparisons have been provided in normal subjects and in dogs on measurements based on $(dp/dt)/P$ derived from the apexcardiogram, using total, instantaneous and developed pressure and polynomial extrapolation techniques of the first and second degree (Willems 1973).

Mean V_{max} indices derived from "total" ACG and LV pressure were almost identical (Table 3). Peak V_{CE} values calculated from the ACG were 5 to 20 units smaller than the ones computed from invasive pressure recordings. Percentage changes of non-invasive and invasive V_{max} indices derived from single beats or from composite data, induced by various inotropic interventions were quite similar. Regression studies resulted in correlation coefficients varying between 0.42 and 0.76.

The indices of $(dp/dt)/P$ derived from the ACG in 51 normal subjects agreed closely with results derived from LV pressure data reported by other investigators.

These results underline the potential usefulness of measurements derived from the first derivative of the apexcardiogram specifically of $(dp/dt)/P$ in a non-invasive pragmatic approach of detecting changes in "myocardial contractility" in man.

Recent results in humans with heart disease, using a newly developed analog device for recording ACG's demonstrated significantly lower normalized first derivatives and peak V_{CE} values in patients with heart failure than in normal subjects (Denef et al 1975).

It may therefore be concluded that indices derived from $(dp/dt)/P$ ratios of the apexcardiogram seem to be promising for the non-invasive evaluation of left ventricular function in humans.

References

1. Abbott J A: The fidelity of the externally recorded human pulse. Am J Med Sci 258, 40 (1969)
2. Agress C M, Fields L G: New method for analyzing heart vibrations. 1. Low frequency vibrations. Am J Cardiol 4, 184 (1959)
3. Bancroft W H Jr, Eddleman E E Jr: Methods and physical characteristics of the kinetocardiographic and apexcardiographic system for recording low-frequency precordial motion. Am Heart J 73, 756 (1967)
4. Benchimol A, Dimond E G: The normal and abnormal apexcardiogram. Its physiologic variation and its relation to intracardiac events. Am J Cardiol 12, 69 (1963)
5. Berry J M: A new method for recording cardiac, hepatic and other pulsations, movements or thrills. Am Heart J 71, 17 (1966)
6. Berson A S, Pipberger H V: Measurment of chest wall vibrations due to the activity of the heart. J Appl Physiol 21, 370 (1966)
7. Brecht K, Boucke H: Neues elektrostatisches Tiefton-Mikrophon und seine Anwendung in der Sphygmographie. Pflügers Arch 256, 43 (1952)
8. Brutsaert D L, Sonnenblick E H: Force-velocity-length-time relations of the contractile elements in heart muscle of the cat. Circ Res 24, 137 (1969)
9. Burger H C, Koopman L J: On the analysis and origin of heart-sounds. Acta Cardiol 4, 131 (1949)
10. Cole J S, Conn R D: Assessment of cardiac apex impulse using fiberoptics. Br Heart J 33, 463 (1971)
11. Cosyns J: Méthodes externes d'appréciation de la contractilité myocardiaque. Acta Cardiol (Brux) 27, 225 (1972)
12. Craige E: Clinical value of apexcardiography. Editorial. Am J Cardiol 28, 118 (1971)
13. Denef B, de Geest H, Kesteloot H: Influence of changes in myocardial contractility on the height and slope of the calibrated apexcardiogram. Am J Cardiol 32, 662 (1973)
14. Denef B, Popeye R, de Geest H, Kesteloot H: On the clinical value of calibrated displacement apexcardiography. Circulation 51, 541 (1975)
15. Denef B, (non published results) 1975
16. Dressler W: Pulsations of the wall of the chest Arch Intern Med 60, 225, 437, 441, 654 and 663 (1937)
17. Dunn F L, Rahm W E: Electrostethography II. New method for study of precordial transmission of cardiodynamics. Am Heart J 44, 95 (1952)
18. Eddleman E E Jr, Willis K, Reeves T J, Harrison T R: The kinetocardiogram. 1. Method of recording precordial movements. Circulation 8, 269 (1953)
19. Edson J N, Flamm G H, Dickes R, Tobin M, Loeb L: Thoracic ballistocardiography. Am Heart J 48, 897 (1954)
20. Falsetti H L, Mates R E, Greene D G, Bunnell I L: V_{max} as an index of contractile state in man Circulation 43, 467 (1971)
21. Fincker J L, Arnold P, Brandt C, Vergnes R, Meyer P: La dérivée première du cardiogramme apexien. Technique et résultats chez les sujets normaux. Arch Mal Coeur 64, 1194 (1971)
22. Foulger J H, Smith P E, Fleming A J: Changes in cardiac vibrational intensity in response to physiologic stress. Am Heart J 34, 507 (1947)
23. Frank O: Prinzipien der graphischen Registrierung. Z Biol 53, 429 (1910)
24. Ghista D N, Brady, A J, Radhakrishnam, S: A 3-dimensional analytical (rheological) model of the human left ventricle in passive-active states. Biophys J 13, 832–854 (1973)
25. Gillam P M S, Deliyannis A A, Mounsey J P D: The left parasternal impulse. Br Heart J 26, 726 (1964)
26. Gleichmann U, Kreuzer H, Loogen F, Wilke K H: Quantitative Messung von Kraft und Geschwindigkeit der Kraftentwicklung im Apexkardiogram. Eine neue Methode. Z Gesamte Exp Med 145, 278 (1968)
27. Groom D, Boone J A: The recording of heart sounds and vibrations. II. The application of an electronic pickup in the graphic recording of subaudible and audible frequencies. Exp Med Surg 14, 255 (1956)
28. Grossman W, Brooks H, Meister S, Sherman H, Dexter L: New technique for determining instantaneous myocardial force-velocity relations in the intact heart. Circ Res 28, 290 (1971)
29. Hartman H: Fonocardiografie en polscurven, in het bijzonder bij aortastenose. Proefschrift Leiden, De Kempeneer, Oegstgeest, 1964.
30. Hollis W J: Time relation of the precordial force-thrust and the acceleration ballistocardiogram from a pendulum and ball bearing platform. Exp Med Surg 16, 116 (1958)
31. Hugenholtz P G, Ellison R C, Urschel C W, Mirsky I, Sonnenblick E H: Myocardial force-velocity relationships in clinical heart disease. Circulation 41, 191 (1970)
32. Ikegaya K, Suzumura N, Funada T: Absolute calibration of phonocardiographic microphones and measurements of chest wall vibration. Med Biol Eng 9, 683–692 (1971)
33. Johnson J M, Siegel W, Blomqvist G: Characteristics of transducers used for recording the apexcardiogram. J Appl Physiol 31, 796–800 (1971)
34. Johnston F D, Overy D C: Vibrations of low frequency over the precordium. Circulation 3, 579 (1951)
35. Kastor J A, Aronow S, Nagle R E, Garber T, Walker H: Air leaks as a source of distortion in apexcardiography. Chest 57, 163 (1970)
36. Kesteloot H, Willems J, van Vollenhoven E: On the physical principles and methodology of mechanocardiography. Acta Cardiol. (Brux) 24, 147 (1969)

37. Kesteloot H, Willems J, Denef B, de Geest H: Experimental study of the height and slope of the calibrated apexcardiogram. (Abstr) Circulation 44, Suppl 2, 32 (1971)
38. Kyle M C, Freis E D: Computer identification of systolic time intervals. Comp Biomed Res 3, 637 (1970)
39. Landes G: Über Brustkorbschwingungen bei der Herzaktion. Dtsch Arch Klin Med 186, 288 (1940)
40. Landes G: Neuere Untersuchungen über Herzstoss und Brustwandpulsation. Dtsch Arch Klin Med 188, 403 (1942)
41. Lohr H A, Vollenhoven E van, Rotterdam A van: Fundamentals in vibrocardiography. Percordial accelerography and acceleration ballistocardiography. Am Heart J 66, 108 (1963)
42. Miller A, White P D: Crystal microphone for pulse wave recording. Am Heart J 21, 504 (1941)
43. Mirsky I, Pasternac A, Ellison R C: General index for the assessment of cardiac function. Am J Cardiol 30, 483 (1972)
44. Motomura M, Omae M, Uehata H, Kumata T, Yaginuma T, Wakabayashi A, Konishi T: An apexcardiographic index (peak $dA/dt)/A$ for the assessment of left ventricular function. Jpn Circ J 37, 1355 (1973)
45. Mounsey J P D: Precordial ballistocardiography. Br Heart J 19, 259 (1957)
46. Pasternac A, Ellison R C, Mirsky I: Assessment of left ventricular function in children by direct and indirect methods. (Abstr), Circulation 44, Suppl 2, 209 (1971)
47. Podolak E, Kinn J B, Westura E E: Biomedical applications of a commercial capacitance transducer. IEEE Trans Biomed Eng 16, 40 (1969)
48. Reale A: Evaluation of the contractile state of the human heart from the first derivative of the apexcardiogram. Circulation 36, 933 (1967)
49. Rosa L M, Luisada A A: Low frequency tracings of precordial displacement and acceleration. Technical comparison of various systems. Am J Cardiol 4, 669 (1959)
50. Schneider H, Klunhaar E W: Precordial low-frequency displacements of the thoracic wall: method of recording and registration. Am Heart J 61, 670 (1961)
51. Sutton G C, Craige E: Quantitation of precordial movement. I. Normal subjects. Circulation 35, 476 (1967)
52. Sutton G C, Prewitt T A, Craige E: Relationship between quantitated precordial movement and left ventricular function. Circulation 41, 179 (1970)
53. Tricot R, Cavailles J, Veber G. Utilisation d'un capteur à variation d'inductance pour l'enrégistrement des phénomènes mécaniques cardiaques. Arch Mal Coeur 59, 805 (1966)
54. Ueda H, Kobayashi T, Sato C, Sakamoto T: Precordial low-frequency vibrocardiography. I. Method of recording. Jpn Heart J 2, 176 (1962)
55. Vas R: Electronic device for physiological kinetic measurements and detection of extraneous bodies. IEEE Trans Biomed Eng 14, 2 (1967)
56. Vetter W R, Sullivan R W, Hyatt K H: Assessment of quantitative apexcardiography. A noninvasive index of left ventricular function. Am J Cardiol 29, 667 (1972)
57. Watson B W, Gay P, Fleming J S: A new instrument for apexcardiography. Cardiovasc Res 2, 108 (1968)
58. Weissler A M, Harris W S, Schoenfeld C D: Systolic time intervals in heart failure in man. Circulation 37, 144 (1968).
59. Willems J, Roelandt J, Kesteloot H: The jugular venous pulse tracing. Proceedings Vth European Congress Cardiology, Vol V Athens Sept 1968, p 433.
60. Willems J, de Geest H, Kesteloot H: On the value of the apex cardiogram for timing intracardiac events. Am J Cardiol 28, 59 (1971)
61. Willems J, de Geest H, Kesteloot H: A new approach to the recording of low frequency precordial vibrations. Acta Cardiol (Brux) 26, 263 (1971)
62. Willems J, Kesteloot H, de Geest H: Influence of acute hemodynamic changes on the apex cardiogram in dogs. Am J Cardiol 29, 504 (1972)
63. Willems J: The normal Apexcardiogram. Ph D Thesis University Leuven. (Arscia ed) Brussels 1973, pp 1–305
64. Willems J: Computer determination of noninvasive measures of "myocardial contractility" from the apexcardiogram. Proceedings Medinfo 1974, North Holland Publ Co, Amsterdam-Oxford-New York 1974, p 687
65. Willems J L, Kyle M C, Pillsbury H C, Freis E D: First derivative of the apexcardiogram and systolic time intervals in the evaluation of myocardial contractility in man. Am J Cardiol 36, 873 (1975)
66. Winter Z, Fabian J: A simple cardiac contractility computer. Med Biol Eng 11, 560 (1973)
67. Zalter R, Hodora H, Luisada A A: Phonocardiography. I. General principles and problems of standardization. Am J Cardiol 4, 3 (1959)

J. L. Willems,
Medical Data Processing and
Cardiology Division,
University Hospital St. Rafael,
35, Capucijnenvoer,
B-3000 Leuven (Belgium)

Adv. cardiovasc. Phys., vol. 5 (Part III), pp. 34–58 (Karger, Basel 1983)

Ballistocardiograms

W. K. Harrison
La Jolla, Calif., USA

Contents

Abstract

Ballistocardiography is concerned with recording of body motion in a way which allows us to determine the cardiovascular force pattern producing it. Genesis of the body motion may be explained by means of equations governing movement of body center of gravity as blood is periodically redistributed during each heartbeat.

Ballistocardiography has been shown by both direct human experimentation and mathematical modelling techniques to provide quantitative information about myocardial contractility. Valuable information on other aspects of cardiac function such as strength and coordination of the heart is also obtainable from it.

Methods of computer analysis of the recordings (using feature recognition and averaging techniques) have been developed to aid the clinician and the research worker in the task of interpretation. A brief history of the method is also presented.

I. Introduction

A. Background

This chapter is concerned with movements of the human body as a whole caused directly by forces of cardiovascular origin. Although these periodic movement patterns share a common genesis, they are astonishingly diverse because of the many different ways and directions in which motion may be recorded. In this chapter, attention will be confined to components of force and acceleration in the head-foot, or longitudinal direction of the body. This aspect of motion has been most thoroughly studied from the scientific and clinical viewpoints.

Elements of a basic system for recording the BCG are illustrated in Fig. 1. The patient is recumbent on a pendulum-like support free to move in his head-foot direction. Although a pendulum hanging from the ceiling is suggested by the diagram, many different varieties of support, or bed, have been employed. It is easy to imagine how such a suspension can undergo periodic movements[1] with each heartbeat of the patient. These movements are sensed by a transducer which converts them into proportional electrical signals of current or voltage. In this case, the transducer is an accelerometer which generates signals proportional to the second time derivative $\ddot{y}$ of the position coordinate y of the support. The patient support, its suspension, and the pickup are collectively referred to as the *ballistocardiograph*. The motion pattern is called a *ballistocardiogram*. Both are commonly designated by the abbreviation *BCG*, a practice which will be followed in this chapter.

The electrical signals from the accelerometer are led to a signal processor, usually an amplifier and filter. These units amplify the relatively small signals from the accelerometer as required to levels necessary for driving the graphic recorder. They may also filter or attenuate noise due to seismic vibration of the building, structural vibration from the BCG itself, and electrical interference such as 50 or 60 Hz pickup from power sources.

Another physiological variable such as an electrocardiogram (ECG), phonocardiogram, or percutaneous carotid arterial blood pressure pulse (CP) is ordinarily recorded along with the BCG. Concurrent physiological signals such as these are quite helpful for accurate interpretation of recordings, particularly those which may come from cardiac patients. In combination with the BCG, the added signal also provides complimentary information on important aspects of cardiac function which neither variable is capable of yielding by itself.

Additional refinements can of course be added to the basic system depicted in Fig. 1. Analog tape recordings can be made for later analysis by computer. Fig. 7 illustrates one

[1] Magnitude of the movements varies, of course, with method of suspension of the patient support. With a pendulum like that illustrated in the diagram, displacements are typically 0.005 to 0.010 cm, and accelerations 1 to 5 cm/s^2. Many readers will appreciate that considerable care is needed to record these small motions for high quality results without undue interference by artifact such as shaking of the building, and various electrical disturbances.

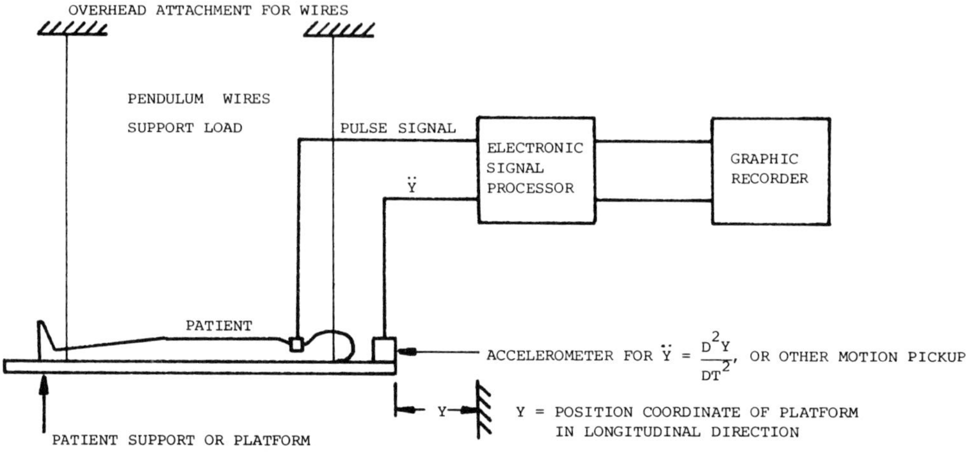

Fig. 1 Schematic diagram of a basic system for recording BCG. The supine patient is recumbent on a light weight pendular platform equipped with an accelerometer to sense second time derivative of its head-foot motion. An amplifier and filter (electronic signal processor) raise signals from the tiny accelerations to levels required for display on a time-base recorder. A percutaneous carotid arterial blood pressure pulse is recorded simultaneously with the BCG to provide valuable complementary information.

variation of this approach. Beside acceleration $\ddot{y}$ of the patient support, other derivatives of displacement y may be recorded. Velocity, the first derivative, and jerk, the third derivative have often been included. Transducers for sensing motion in orthogonal directions to y have sometimes been mounted on the platform.

Fig. 2 illustrates acceleration (A), velocity (V) and displacement (D) BCG typical of those commonly seen in healthy young men and women free of clinical heart disease. Patterns are representative of those obtained from "ultra low frequency" (ulf) or high fidelity apparatus described later. Timing relationships with BCG, heart sounds, and phases of the cardiac cycle are also indicated. The acceleration record, to which this chapter is devoted, is characterized by the prominent IJ wave inscribed in the first half of the ejection period of the ventricles. This wave is closely related to initial acceleration of blood from the left ventricle into the aorta, and to a much lesser extent to corresponding events in the pulmonary circulation. The waves labeled L_0, L, M, which occur at the time of closure of the semi-lunar valves, are related to the aortic blood accelerations involved in closure of these valves. Wave L_0 is sometimes denoted by K. The N wave, inscribed during the initial rush of blood through the atrioventricular valves into the ventricles, is correlated with the magnitude of these blood accelerations. Further filling of the ventricles by the atria and myocardial motion are responsible for the initial F – G – H waves of the BCG. See Scarborough et al (1958) for detailed discussion of these waves and additional references.

This orderly acceleration pattern in healthy people usually shows minor variation from heartbeat to heartbeat throughout the respiratory cycle in eupnea. Variation is largely abolished during apnea if the Valsalva maneuver is avoided, and the typical resting pattern of each subject may change in response to various stresses such as exercise, drugs, smoking, etc. Amplitudes of the waves are highly responsive to increases in blood acceleration as would be expected in exercise or interventions which affect the inotropic state of the myocardium. A great deal is known about the genesis of BCG of healthy subjects in terms of fundamental physiological variables such as pressure and blood flow. Some of these facts will be reviewed later in this chapter.

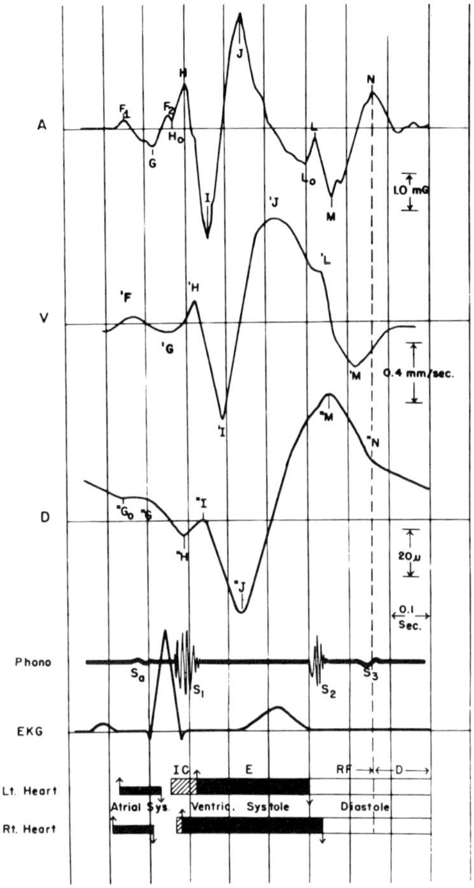

Fig. 2 Normal human displacement (D), velocity (V) and acceleration (A) ballistocardiograms from an ultra-low frequency (ULF) system and their temporal relation to events in the cardiac cycle. The BCG, heart sounds (phono) and ECG tracings were based on the measurements from 25 males, 20–29 years of age. The block diagrams at the bottom represent the timing of cardiac events on the two sides of the heart and are based on data supplied by Braunwald and by Wiggers (see text). IC = isometric contraction (cross-hatched), E = ejection phase (solid black), RF = rapid filling, and D = distasis.

Calibrations for the three ballistocardiograms are shown on the right. mG represents milli-G; 1.0 mG (0.001G) is approximately equal to 1.0 cm/s^2; μ represents micron; 20 μ is equal to 0.02 mm. The BCG tracings represent external motions of the body; headward motions are upwards. The tips of displacement waves should coincide in time with the tips of corresponding acceleration waves; however, the displacement waves are slightly advanced in time owing to the presence of a small phase lead.

From Scarborough et al: The nature of records from ultra-low frequency ballistocardiographic systems and their relation to circulatory events. Am J Cardiol 2, 613–641 (1958). Courtesy of the authors and publisher.

In diseases of the heart, in general, the stereotype normal pattern of BCG is replaced by widely differing waveforms. Fig. 8 illustrates "abnormal" as well as "normal" patterns. There is usually much greater heartbeat to heartbeat variability, both in magnitude of waves and in their pattern. This is not surprising in view of the many different possibilities which exist for aberrant cardiac function. With a few exceptions, there is no exact correspondence between particular anatomic lesions and characteristic BCG patterns. Relatively little is known about the genesis of abnormal BCG patterns, and about their interpretation in terms of compromised cardiac function in disease. This situation is the direct result of a corresponding lack of knowledge of circulatory phenomena such as heart motion and phasic arterial blood flow in cardiac disease.

B. History of BCG

Man's first consciousness of movement of his body in response to the heartbeat must have come quite early in the evolutionary process. Perhaps it came about during a rest period just after arduous pursuit of food or enemy, while the heart still beat strongly. Few of us have not, at one time or another, lay quietly on a mattress which happened to be sprung so that our body motions resonated with the heartbeat. In the process of weighting one's self on some kinds of scales, motion of the pointer in time with the heartbeat is very noticeable. In fact, the first published recordings of the BCG by Gordon (1877) were derived from just such an instrument. Gordon went much further than merely magnifying pointer motion of the scale and recording it on a smoked drum Kymograph. He describes a pendular bed for the horizontal subject which might have compared favorably with ultra low frequency BCG's presently in use. Unfortunately, he did not elaborate on his experiments, and published only two selected patterns obtained from the scales and bed in his short article.

Closely following Gordon's paper in the same volume of the *Journal of Anatomy and Physiology* is a brief but discerning note by Trotter (1877). In it he discusses the body movements reported by Gordon in terms of the fixed position of the center of gravity of a suspended body acted on purely by internal forces. This concept was later employed to advantage both by Henderson, who may not have known of Trotter's article, and by Burger. Trotter also considered the practical matter of getting accurate tracings of body motion, and recommended coupling the tracing pointer to a rod held firmly between the subject's teeth. Although this suggestion would appear to have merit for investigations requiring an accurate sampling of skeletal motion, the author knows of no instance of its use in any of a number of such research studies which have since been performed. In 1880, a few years after Gordon's and Trotter's scientific reports, Landois (1880) published clinical recordings of body motion of normal subjects and those of a patient with aortic insufficiency.

Apparently work on the method was dropped for the next 25 years, since no reports have been found until Yandell Henderson's extensive article (1905). Henderson recorded displacement of a suspended, pendular bedtype support for supine subjects much like the modern ultra low frequency apparatus. Numerous calibrated displacement recordings were published to illustrate inter-individual variability of the pattern. Henderson realized the inherent problems in use of the recordings to estimate stroke volume. Nevertheless, eight years later with Douglas, Haldane, and Schneider (1913), he employed a crude variation of the method to conclude incorrectly that cardiac output was not altered by altitude change from sea level to the top of Pike's Peak.

For 23 years after the Pike's Peak expedition, the BCG received very little attention. The physician Satterthwaite (1913) again noted swinging of the pointer of his office scales as he weighed patients, and made recordings of the movement. Heald and Tucker (1922) appear to have been the first to use an electrical method for recording the BCG, though their procedure could not distinguish headward from footward motions of a frame supporting the standing subject. By use of a seismograph mounted on a table supporting a supine subject, Angenheister and Lau (1925) made recordings which resemble more modern BCG. In 1933, the Swedish physiologist Abramson constructed a chair type apparatus for recording BCG of seated subjects. None of these investigators persisted in his experiments, and the BCG remained an obscure curiosity. It was poorly understood by physicists, physiologists, and physicians, if indeed they even knew of its existence.

Isaac Starr's attention was directed to the method in 1936 after he had worked for years on improvements in techniques for measurement of cardiac output. Starr was critical of existing procedures, especially their unsuitability as clinical tools, and recognized potential advantages of the BCG for estimation of cardiac output in patients. His first major report (1939), proposed formulae for stroke volume based on measurements of calibrated systolic waves of BCG recorded on a stiffly sprung bed apparatus which he described in detail. This initial work really ushered in the rapidly expanding era of investigation of cardiovascular dynamics which continues today. Starr's scholarly, voluminous research is without question the single most important source of information about the clinical and physiological aspects of the BCG in existence. In 1967, with the collaboration of the physicist Noordergraf, he published a comprehensive review of the field including developments up to that time. This book should be read by anyone wishing to work with the BCG or interpret the recordings.

In the 10 years from mid World War II to the early 1950's, BCG research was characterized by empirical clinical applications and a regrettable controversy about instrumentation. Electronic recording techniques were enthusiastically and uncritically embraced, while users seemed to loose sight of the need for a better fundamental understanding of BCG. There was wide disagreement about exactly what physiological events were causing observed effects and how best to relate obtainable information to clinical knowledge as a whole. Several competing types of BCG apparatus emerged (Nickerson 1944, Dock 1949), and it soon became clear that recordings made on one type were not comparable to those from another. This confusing state of affairs disenchanted most clinical users at the same time that interest in new, catheter-based invasive procedures was rapidly increasing. The result was a virtual cessation in clinical investigation utilizing BCG by the early 1960's.

It is not surprising that yield of useful medical results from the BCG in this period was low. One significant relationship did emerge, however, and that is the identification of an abnormal BCG pattern (Fig. 8) as a risk factor for coronary heart disease in males. This relationship was demonstrated to hold irrespective of which form of BCG instrument was employed in independent studies by Starr (1961), Baker (1967) and Kiessling (1970). Baker's study separated effects of BCG grade and age of subject by means of multivariate statistical analysis, with results illustrated in Table 1. This shows that age and BCG grade are about equally important for prognosis of a coronary event in asymptomatic normal subjects. In patients who had already had a heart attack or other manifestations of the disease, age was a much weaker predictor of an additional future event than was abnormality of the BCG. The results of Baker's study and those of the other investigators were obtained in spite of major defficiencies of the apparatus they employed and many drawbacks in the methods for rating the BCG which existed at the time.

Problems with instrumentation were resolved by the entry of physicists and engineers into the field. Burger (1953) in Europe, Talbot and Harrison (1955) in the United States, applied techniques of mathematical modelling to the patient-BCG system to demonstrate desirable characteristics for optimal recording. These investigations and others established the so called "ultra-low frequency" (ulf) system with recording of acceleration of the bed in head-foot direction. A committee headed by Cunningham (1970) published desirable standards for the ulf system which ensure comparable recordings. Practical ulf BCG beds suitable for clinical use were described by Tolles (1966), Harrison and Talbot (1966), and Reeves (1957). A manufacturer[2] currently produces an air bearing BCG based on the design of Harrison and Talbot which is suitable for research and clinical use.

Significant progress was also made in advancing understanding of the BCG in terms of the causative circulatory physiological events. Specific features of the more important studies are discussed in the next section of this chapter. Noordergraaf (1961) and his students explored the BCG extensively using an electrical analog

A. **Risk of event within nine years: Males, Pure Normal Group**

Age	*Starr form* 1	*2*	*3*	*4*
30	0·025	0·045	0·080	0·137
40	0·041	0·072	0·125	0·206
50	0·066	0·114	0·189	0·299
60	0·103	0·174	0·277	0·411
70	0·160	0·256	0·386	0·467

B. **Risk of recurrent event within nine years: Males, IHD Group**

Age	*Starr form* 1	*2*	*3*	*4*
30	0·344	0·500	0·656	0·784
40	0·350	0·506	0·662	0·789
50	0·356	0·513	0·668	0·793
60	0·362	0·520	0·674	0·797
70	0·369	0·526	0·680	0·802

Table 1 A. These estimates of risk of cardiac event (myocardial infarction, symptomatic ischemic heart disease) were based on observations of 273 healthy males followed for 9 years. The method of rating BCG 1, 2, 3, or 4, according to increasing abnormality is described in section IV.A. of text. Note that both age and BCG grade have a strong influence on predicted risk.

B. These estimates of risk of a recurrent cardiac event were based on data from 259 men who had unequivocal evidence of ischemic heart disease. Note that age has relatively little bearing on predicted risk in this group. Instead, there is strong dependence of risk on BCG grade. (From Baker et al (1967), with their permission and that of the publisher.)

[2] Royal Medical Corporation, 110 Wynn Drive, Huntsville, Alabama 35806, U.S.A.

of the arterial (1963) and later the pulmonary (1968) circulations. The most recent and comprehensive study of the relationship between BCG and aortic flow derived from this electrical analog is the thesis by Verdouw (1970). Harrison also developed a mathematical modelling approach by which cardiovascular forces responsible for the head foot BCG were calculated from aortic, pulmonic, intra cardiac flow patterns and cardiac mass motion. Baan (1972) developed a rationale for estimating the correlation between initial maximum aortic blood acceleration and the maximum slope of the HI wave of the BCG. He tested the relationship in human subjects by recording simultaneous BCG and aortic blood acceleration in humans undergoing cardiac catheterization. These results demonstrated the validity of the ulf BCG as a noninvasive index of cardiac function, and supported earlier work by Harrison et al (1969) relating BCG to maximum left ventricular pressure derivative in man.

As physiological and instrumentational questions were in the process of resolution, research on BCG in the last five years has taken new directions. A small number of clinical correlations has been undertaken, this time with full knowledge by the investigators of strengths and limitations of the technique. Jackson's studies of aortic valve disease and hyperkinetic ventricular action (1971), and Smith's application (1968) to monitoring after cardiac transplantation are examples. Inadequacy in our ability to interpret the BCG record, at least in a clinically useful way, has led to development of methods for analysis of the heartbeat patterns by digital computer. Eddleman's group, with extensive experience in computer analysis of many cardiovascular recordings, has perfected apparatus and programs for detailed on and off line measurements of timing, amplitude, and pattern characteristics (Bancroft, 1973). Their approach, and Harrison's procedure (1972) employing a mini-computer, rely on interaction between the computer operator and the program for sampling of data and checking. A later procedure, developed by Harrison (1974), relieves the operator of intimate involvement with function of the computer program, which operates in an IBM 370/145 computer system. These new approaches to computer reading of BCG are much superior to the subjective methods used in past research. Because of this it seems likely that they will be applied to carefully selected clinical and epidemiological problems in future years.

II. Genesis of BCG and Hemodynamic Correlations

A. Movements of Center of Gravity of Blood in Genesis of BCG

The very first investigators of BCG speculated about its genesis. Gordon (1877), Trotter (1877), and Henderson (1905) understood the phenomenon in general terms of periodic movement of the center of gravity of the patient as blood was redistributed with each heartbeat. While further attempts to utilize this viewpoint were made by others, Burger (1953) and Noordergraaf (1956) were the first to develop the concept in precise physical terms. They used the fact that motion of the center of gravity of a freely suspended body referred to its skeleton, and motion of the skeleton itself are reciprocal.

Noordergraaf (1956) took advantage of this simple relationship to calculate BCG from published arterial pressure pulse contours, arterial wall properties, and anatomical measurements. His reasoning goes as follows for the longitudinal or head-foot ulf BCG: The body is considered to be divided up into cardiovascular and non-cardiovascular mass increments. The cardiovascular increments, $m_i(t)$, are time variant, while the remaining masses m_j in the body are not. $m_i(t)$ and m_j are located with respect to the skeleton by corresponding longitudinal coordinates y_i. The sum of all $m_i(t)$ and m_j equal the total body mass m_s. The masses m_i (t) vary periodically throughout the cardiac cycle, with a minimum value at end diastole for the arterial components. The center of gravity of the body is then defined as

$$y_c(t) = \frac{\sum_i m_i(t) y_i + \sum_j m_j y_j}{m_s}.$$

In this equation, advantage may be taken of the fact that the left term of the right hand side is time varying above a definite minimum level. This level, along with the other terms representing non-cardiovascular tissue can be regarded as a constant which may be arbitrarily equated to zero by choice of reference axis. If

ς represents blood density, and Δv_i pulsatile blood volume of each vascular segment, the preceding equation reduces to

$$m_s y_c(t) = \varsigma \sum_i \Delta v_i(t) y_i .$$

Noordergraaf employed results from the physiology literature to calculate Δv_i. For the arterial pulsations, it was necessary to select vessel segment minimum volume v_{id}, wall thickness to radius ratio a_i, and wall modulus of elasticity E in addition to the pressure pulse pattern $p(t)$. Wall material was assumed to be incompressible, and motion of tissue adjacent to the arteries was neglected. The relation between $\Delta v_i(t)$ and $p(t)$ was based on Hook's law (linear elasticity):

$$\Delta v_i(t) = \frac{3v_{id}(a_i + 1)^2 p_i(t)}{E(2a_i + 1)} .$$

These calculations of "excess blood volume" $\Delta v_i(t)$ were made on 115 arterial segments, each so short that the pulse wave passed in 10 milliseconds. The arterial component of the ulf displacement BCG was simply the reciprocal pattern: minus $y_c(t)$. The velocity and acceleration contributions were calculated by successive numerical differentiation.

This procedure was ultimately applied not only to the main arteries, but to the heart chambers, the pulmonary circulation, and smaller arterial segments than the original 115. Contributions from each source were then summed at each instant of time in the cardiac cycle to produce the displacement BCG. This pattern, and its successive derivatives, closely resemble those recorded from normal, healthy humans. Similarity of the derived and measured BCG support the rationale employed for the derivation and the inevitable simplifying assumptions which are necessary. The acceleration or force BCG Noordergraaf obtained after taking into consideration all but venous contributions is quite similar to that in Fig. 2.

Mathematical formulation of Noordergraaf's procedure indicates that the BCG pattern at all times reflects contributions from every part of the circulation. For this reason, any particular pattern feature such as depth of I wave, height of N wave, etc., may not be regarded as being caused solely by a single, exclusive, anatomically localized event. This realization must always be kept in mind when interpretations are placed on the BCG pattern. However, reservations should be applied in a way which recognizes the fact that some features of BCG are caused predominantly by particular hemodynamic events, with insignificant contribution from other sites in the circulation. Examples of these hemodynamic features which may be important for future clinical use are initial aortic blood acceleration in flow across the mitral and aortic valves. Additional theoretical studies of the genesis of the BCG considered in the next section furnish more guidance on the extent to which its features may be expected to respond primarily to hemodynamic events such as these.

B. Phasic Regional Blood Flow in Genesis of BCG

Movements of center of gravity of blood and body in response to heart action are easily formulated for analytical study of the genesis of BCG. Motion of center of gravity is a valid physical concept, but an unfamiliar one to physiologists and clinicians who most often think in terms of easily measured blood pressure. Also, it has not been possible to measure position of center of gravity of blood independently of the BCG. Theoretical predictions derived from this concept can thus not be verified in direct animal or human experimentation.

With the preceding facts in mind, Verdouw (1970) and the author (1970) sought independently to derive body accelerations analytically from phasic blood flow patterns. The aortic root flow pulse is a much more familiar quantity than motion of center of gravity of blood, and it is receiving increasing attention in cardiovascular research. Initial acceleration of blood in the ascending aorta has been shown to be a good indicator of myocardial contractility. It is relatively insensitive to variations in myocardial preload and afterload as found in the circulation of free living, healthy humans.

Verdouw and Noordergraaf (1970) published the most detailed study of relationships between ventricular ejection flow and BCG. Their approach used an arterial "transfer function" derived from an elaborate electrical analog of the systemic arterial system. The analog had been extensively tested and found to be accurate for pressure and flow components up to 20 Hz. The transfer function and its first and second derivatives could be convoluted with aortic valvular flow to produce displacement, velocity, and acceleration ballistocardiograms due solely to blood flow in the systemic arterial system. Effects due to cardiovascular activity other than arterial flow were absent from this simulation. However, omission of these features was not a handicap in correlation of specific features of aortic flow pulse and BCG since effects of the other cardiovascular activity such as pulmonic blood flow and heart motion are additive.

A variety of aortic root blood flow patterns based on triangular approximations which are similar in shape to the physiological pattern were arranged to explore the effects of varying initial flow acceleration, peak flow, stroke volume, and terminal retrograde flow. These systematic changes in contour of flow pulse made it possible to identify corresponding effects in the BCG such as deepening of waves or changes in their timing. Triangular and other flow patterns constructed by joining straight line segments were adequate representatives of normal human physiologic flows. A later report (1973) by the same authors illustrated features in the flow pattern thought to be appropriate for diseased hearts and their effects on the calculated BCG.

Verdouw computed BCG patterns corresponding to four left ventricular outflow curves with succesively decreasing initial acceleration. These triangular approximations had equal area or stroke volume, peak flow value, and ejection time. Depth of I wave below zero level in the BCG showed a decrease which was linear with initial flow acceleration. Height of J wave also decreased, but to a much smaller degree. There were minimal changes late in the ejection period and immediately thereafter.

A similar procedure explored effects of varying the terminal slope of the flow pulse while keeping initial acceleration and peak ejection flow unchanged. In this simulation, depth of I wave was the same in each case, and only negligible changes in height of J were found. Not surprisingly, the most noticible differences were apparent in late ejection and early diastole. K and M waves became progressively deeper as steepness of the terminal slope of the flow pulse increased.

Verdouw also tested for effects on the BCG caused by varying peak flow while keeping constant initial and terminal slopes of the triangular patterns. Area under the flow curve, or stroke volume, varied in this series of experiments. Height of J wave tip above zero baseline increased linearly with peak flow, while depth of I wave was unaffected. There were also timing differences in L – M – N waves of early diastole corresponding to varying ejection period of the flow patterns.

Additional calculations were made for ventricular flow pulses with 4 and 5 discrete slopes to more closely approximate the physiological pattern shape. BCG corresponding to each pattern always showed differences which were related to the changes. Variations in shape of the peak of the flow produced alterations in the neighborhood of the J wave tip. Retrograde flow after the ejection period accentuated the early diastolic L–M waves.

Verdouw's studies clearly showed that initial acceleration of ventricular outflow was strongly reflected in depth of I wave below zero acceleration level of arterial BCG. This demonstration linked the so called "initial ventricular impulse," an accepted index of myocardial contractility, with a specific feature of the easily recorded BCG. Verdouw and Noordergraaf did not examine the maximum HI slope, or jerk. If they had, it is highly likely that this quantity also would have exhibited a similar relationship to flow acceleration as depth of I wave. Baan et al (1972) found this relationship both experimentally and theoretically. HI jerk and I wave amplitude are known to be correlated with each other, at least in BCG patterns which resemble the normal stereotype.

It is reasonable to suppose that the strong linear relationship found in these model studies would be weakened somewhat in practice. Effects of heart movement, right ventricular ejection, and venous flow were left out of the Verdouw simulation, but are superimposed on the arterial BCG in real life. These effects are smaller than the arterial features, at least in the normal circulation, as the author's (1970) independent analysis showed. This investigation used a different approach from that of Verdouw and Noordergraaf, but largely supports their findings. In the author's results, height of N wave was more dependent on vigor of ventricular filling in early diastole than on purely arterial effects which produced this wave in the Verdouw simulation. It was also apparent that abnormalities of ventricular motion, known to occur in myocardial infarction and other disease states, could themselves produce abnormal BCG. This was especially likely to happen if these effects were additive to an arterial BCG of small size which is not uncommon in disease.

These independent model studies support interpretation of the BCG from normal persons in terms of myocardial contractility. Caution is called for when reviewing BCG from abnormal or diseased hearts, because in these situations hemodynamics and cardiac motion may vary considerably from examples tested in the model studies. More research is required both in humans and on models to explore fully these effects so that BCG from all kinds of cardiac patients may be interpreted in a more sensitive and knowledgeable way.

III. Instrumentation and BCG Test Procedure

A variety of different instrumentation for recording BCG has been employed in the past. Of these, the so called ultra low frequency (ulf) apparatus possesses the best characteristics for registration of cardiovascular forces with minimum distortion. Current research and clinical activities employ the ulf BCG almost exclusively. This section deals with the force or acceleration ulf BCG, although mathematical modelling principles employed in its analysis are equally applicable to the other systems. For a detailed discussion of the other types of apparatus which have been developed, the reader should refer to the appropriate chapters in Starr's book (1967), or the article by Talbot (1955).

A. Mathematical Modelling of BCG Patient and Apparatus

An essential step in analysis of a mechanical system such as the BCG is mathematical modelling of the patient in an appropriate way. Mechanical properties of the human body which affect its vibrational behavior in general are sufficiently complicated to prevent more than token recognition in a model. Spatial variation of tissue mass, elasticity, viscocity, and geometrical factors are examples of properties which present models either ignore or include only in a grossly simplified way. Neglect of properties of real human bodies such as these in mathematical modelling of the BCG system is justifiable retrospectively when predictions of vibrational response obtained with the simplified model are compared with experimental data from real systems. Only gross characteristics of mass, viscosity, and elasticity have been incorporated in human body vibrational modelling of the BCG up to now. Efforts devoted to greater refinement have probably been discouraged both by rapidly escalating mathematical complication and a fair degree of experimental justification for the simplified approach. Even a highly simplified model has real value in advancing our understanding of the actual BCG system.

A "rigid mass" model of the body has usually been adopted by investigators of the dynamics of BCG. The mass (m_b) representing the body is considered to be coupled to its support by elastic and viscous elements (k_b, r_b) in parallel. This analogy to a supine human is illustrated in Fig. 3 a. Here, mass of the model is usually taken to be total mass of the patient, a choice which allows nothing for weight of those dorsal tissues which supply the elasticity and damping. Such an omission is of little consequence to appropriateness of the concept of the model, though it may be important if detailed comparisons are to be made between results from the model and those from experiments with actual human bodies.

There is considerable similarity between transient longitudinal displacement of a fresh, supine cadaver tapped gently on the top of the head, and that of the damped simple harmonic oscillator model, acted upon by an impulse. If head-foot displacement of a living body driven by the natural heartbeat is recorded, as illustrated in Fig. 3b, the same kind of partially damped, oscillatory pattern results. The longitudinal vibrational mode of most supine humans has its damped fundamental natural frequency in the narrow range of 3–6 Hz. As in the well known response of the damped, single degree of freedom spring-mass-damper system, there is appreciable magnification of motion caused by applied force components near resonance, and strong attenuation above it. (See, for example, Freberg 1949.) Because the spectrum of cardiovascular force in either health or disease contains significant components in this band, direct recordings of simple head-foot body motion of the supine patient may differ markedly from the applied force pattern. It is curious that early clinical research using the so-called "direct body" BCG method was for many years carried out without knowledge of these facts and their implications for results obtained.

If motion in directions other than head-foot is considered, a single rigid mass approximation

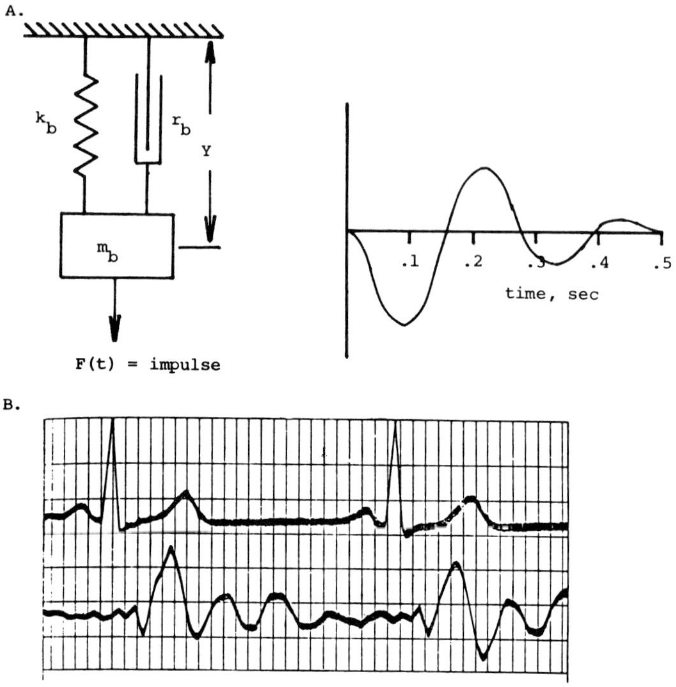

Fig. 3 a. Most approaches to mathematical modelling of the BCG have employed the concept of a "rigid mass" approximation for the human body. The mass is considered to represent a supine subject coupled to his surroundings by means of visco-elastic dorsal tissues represented by parameters r_b and k_b. Coordinate y represents displacement of skeleton, and impuse $F(t)$ a sharp tap on the top of the head which causes an initial footward excursion and quickly damped oscillation as indicated in the graph. Fundamental frequency of most bodies in this vibrational mode is in the range 3–6 Hz.

b. Actual recording of displacement y of a supine human shows an initial footward displacement at the beginning of ventricular ejection and subsequent transient oscillations indicating response to a complex periodic force pattern. These so called "direct body" BCG recordings were used in empirical clinical investigations. Upper tracing is the subject's electrocardiogram.

for the body may be inappropriate. Even if it were a good representation, modelling of the body would require at least six degrees of freedom, three translational and three rotational components. This approach was taken by Talbot (1958), but he did not develop it in detail to obtain new results which might advance our understanding of BCG in general and focus attention on further applications. In fact, measurements of body properties or experimental recordings of body motion which would permit this still do not exist. Consequently, almost all of our present knowledge of the BCG as a noninvasive index of cardiac function is based exclusively on recordings of motion in the head-foot direction.

The first resonance of the supine human vibrating in longitudinal mode lies squarely in the frequency band of pulsatile blood flow (Patel et al 1965). To improve on this undesirable situation for accurate physiological recording, the body support in a BCG apparatus must be designed so that combined body and support system resonances are moved outside the limits of the cardiovascular force signal. This is done by cradling the patient (m_b) on a pendulum-like support (m_s) which is very loosely coupled to its surroundings. Addition of the support, as indicated in Fig. 4, results in a two degree of freedom system in terms of the mathematical model. Coupling of the patient to his support (r_b, k_b) is made as tight as possible consistent with comfort. Equations of motion of the two degree of freedom system of patient and support in terms of variables defined in Fig. 4 are:

$$m_b\ddot{y}_b + r_b(\dot{y}_b - \dot{y}_s) \div k_b(y_b - y_s) = F_b \qquad (1)$$

$$m_s\ddot{y}_s + r_s\dot{y}_s + k_s y_s + r_b(\dot{y}_s - \dot{y}_b) + k_b(y_s - y_b) = 0 \qquad (2)$$

in which y_b and y_s are displacement of body and of support, respectively, and F_b is the cardiovascular force pattern caused by heart

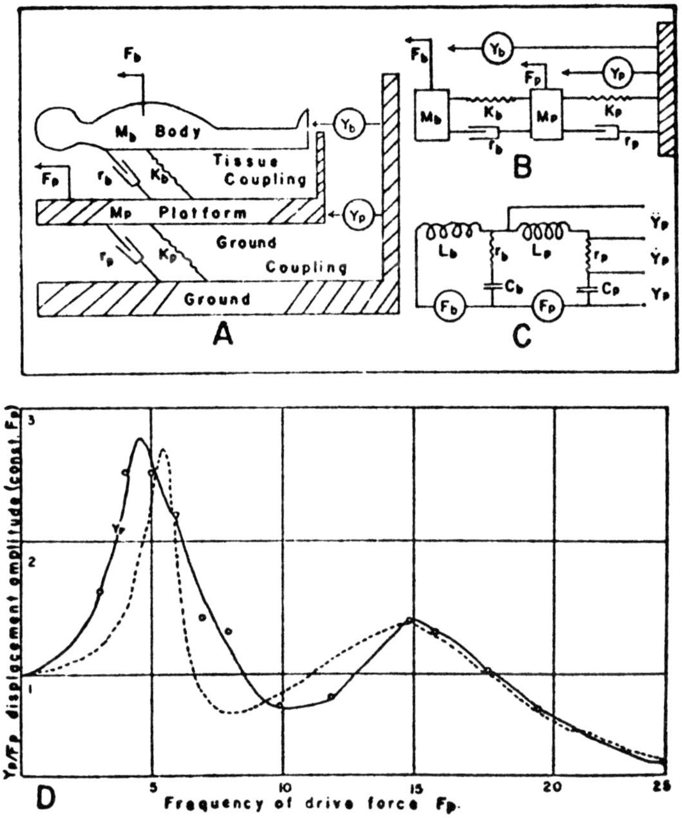

Fig. 4. Mechanical analog of the body-bed ballistocardiograph system. A. Distributed tissue elasticity (dorsum and feet) lumped in k_b. Distributed tissue viscosity (shear and compression) lumped in r_b. Distributed mass assumed lumped as shown.
B. Components rearranged to show derivation of exact electrical analog, C, regarded as a computer, wherein driving force (voltage) is applied to either body or platform, and $y_p = D\ y_p = V\ y_p = A$ are read as voltages.
C. Solid line: (experimental) response of platform y_p when shaken by F_p through ballistocardiograph spectrum (1 to 25 cycles per second). Dotted line: (theoretical) response computed by C for mechanical analog of body and bed, A or B. From: Talbot S A, Harrison W K, Jr. Dynamic comparison of current ballistocardiographic methods, Part II: Effect of a platform in ballistocardiographic dynamics. Circulation 12, 5 (1955). (reprinted with permission) Note that k_p and r_p of the figure correspond to k_s and r_s of text. Also, D does not represent results from a system with optimal characteristics such as that described in text.

and blood acceleration in the body of the subject as indicated in Fig. 4. Parameters of the model are chosen so that acceleration of the support $\ddot{y}_s$, which is measured in practical systems, closely approximates force pattern F_b.

To illustrate the desired relationship between cardiovascular force F_b and acceleration of the support y_s in terms of this rigid mass mathematical model, Eqs. (1) and (2) may be divided by F_b:

$$\frac{m_b \ddot{y}_b}{F_b} + \frac{r_b(\dot{y}_b - \dot{y}_s)}{F_b} + \frac{k_b(y_b - y_s)}{F_b} = 1 \quad (3)$$

$$\frac{m_s \ddot{y}_s}{F_b} + \frac{r_s \dot{y}_s}{F_b} + \frac{k_s y_s}{F_b} + \frac{r_b(\dot{y}_s - \dot{y}_b)}{F_b} + \frac{k_b(y_s - y_b)}{F_b} = 0 \quad (4)$$

If $\frac{r_s \dot{y}_s}{F_b}$ and $\frac{k_s y_s}{F_b} \langle\langle 1$, restoring forces of the support are much less than cardiovascular forces[3], and (4) becomes approximately

$$\frac{m_s \ddot{y}_s}{F_b} + \frac{r_b(\dot{y}_s - \dot{y}_b)}{F_b} + \frac{k_b(y_s - y_b)}{F_b} = 0 \quad (5)$$

Add (3) to (5) to obtain

$$\frac{m_s \ddot{y}_s}{F_b} + \frac{m_b \ddot{y}_b}{F_b} = 1 \quad (6)$$

[3] Order of magnitude of cardiovascular forces in the resting human is typically 100 gm. Eg., two stroke volumes of 50 gm expelled with accelerations of 1 g. Thus, $k_s y_s$, the spring force developed by the support, should be around 1 gm, and damping force $r_s y_s$, if present, should be no more.

From (6), the desired simple relationship between measured acceleration $\ddot{y}_s$(the BCG) and F_b,

$$\ddot{y}_s = \frac{F_b}{m_s + m_b}$$

results if $\ddot{y}_b = \ddot{y}_s(1 + \ddot{a})$, and $\ddot{a}$, the "error" term, is small compared to unity. It can be shown by substitution in (4) that $\ddot{a}$ will indeed be negligible if k_s/k_b and $r_s/r_b \ll 1$, and frequency $\omega/2\pi$ of applied force F_b lies in the range $\frac{k_s}{m_b} \leq \omega^2 \leq \frac{k_b}{m_s}$. This range will be greatest with tight coupling of support mass m_s to body mass m_b, and the lowest practical ratio of m_s to m_b.

Detailed specifications have been given by Cunningham (1970) for optimal values of these and other parameters of ulf systems. Cunningham's report is based on recommendations by committees of the United States and the European BCG Research Societies. All prospective users of the method should be familiar with these specifications before setting up their laboratories.

B. Ulf BCG Apparatus

Many variations of ulf BCG apparatus have been described since the advantages of a long period or aperiodic suspension as described in the preceding section, first began to receive recognition. Designs intended for clinical or research use may be classified as either mechanical spring types, or air bearing suspensions, according to the method of anchoring the patient support or bed (m_s). Completely aperiodic systems have also been tried, but these were purely exploratory experiments which were unsuitable for routine use. In one such device reported by Talbot et al (1954), the patient rested in an 11 pound boat which floated in a pool of mercury. The head-foot acceleration patterns recorded from healthy subjects on this apparatus were indistinguishable from those obtained from the later mechanical or air bearing supports which met desired performance standards.

Another unusual aperiodic system due to Hixon and Beischer (1964) was developed in the United States space flight program in the early 1960's. The rationale for these experiments doubtless recognized the ease with which acceleration BCG might be taken for cardiovascular monitoring of astronauts in weightless conditions of space or orbital flight. Such conditions were simulated in a four engine jet transport aircraft flying parabolic maneuvers in which weightlessness or "free fall" conditions were maintained for periods of approximately 45 seconds. For these tests, a specialized constraint platform was moulded to the dorsal contours of the experimental subject, who was in addition snugly attached to it with multiple straps. Linear accelerations of this platform in three mutually perpendicular directions, an angular acceleration, and ECG were telemetered to recording equipment carried aboard the aircraft. The subjects' longitudinal acceleration recorded in these tests was compared to his headfood BCG obtained on an air bearing bed in the laboratory and found to be almost identical. It is regrettable that this technique was not subsequently employed in the space program. However, results from it will be useful for validation of three dimensional BCG taken on earth if appropriate apparatus can be developed for this purpose at some future time.

Mechanically sprung BCG are usually derived from upright or inverted pendula. The simplest arrangement consists of a lightweight (less than 5 kg) platform suspended by parallel wires from an overhead support. Wires must be about 3 meters long to provide the desired low frequency response. Inclination of the support wires, or even attachment to a common axis does not significantly reduce the need for a room with unusually high ceiling in which to set up this form of BCG. Because of such a major drawback, this conceptually simple system has not been frequently employed.

A modification of the suspended pendulum design which allows a shortened pendulum length has been used much more often. In this version, the patient support is displaced laterally by a pair of sharply pointed pins, cradled in sockets, one at each end. These pins act as a destabilizing influence on the main pendular support so that its fundamental frequency is lowered as much as desired. Gordon (1877), and later Henderson (1905) used this principle of suspension in their early exploratory apparatus. Reeves et al (1957), have described a more modern version. A commercial BCG of this type with overhead cables about 2 meters long was also manufactured by Evershed-Enraf Company of Delft, Holland.

Apparatus based on inverted pendulums probably received attention because they allowed designers to eliminate the long suspension wires and super structure associated with them. In such a relatively compact apparatus, it might be justifiably claimed that an acceptable degree of clinical practicality had been obtained. Tolles (1966) reported on a BCG bed using inverted spring pendula which were destabilized by coil springs so that the desired low fundamental frequency could be set for each patient. A commercial BCG using inverted spring pendula was also offered by the Schwarzer Company of Munich, Germany. Talbot et al (1958) reported on a BCG bed design employing a combination of positive and negative gravitational pendulums. This arrangement had the advantage that no adjustments for frequency had to be made to allow for variation in weight of the patient as in the Tolles and Schwarzer apparatus.

Use of an air bearing in a BCG bed design was first reported by Nyboer (1957). Shortly afterward, a commercial ulf BCG using a single spherical air bearing centrally located under the patient was introduced by Cordis Corporation of Miami Florida. About the same time, Cunningham (1961) described an apparatus using three small pad-type air bearings to support a rigid, light weight platform constructed of aluminum honeycomb. Both the Cordis and Cunningham designs had three degrees of freedom: transverse, longitudinal, and rotation in yaw. Somewhat later, Harrison and Talbot (1966) reported on another variation of the air bearing principle which was well suited for use on a flouroscope for recording BCG during cardiac catheterization. In this design, the patient was supported by a pallette of styrofoam and aluminum which floated on an air film provided by an industrial vacuum cleaner (Fig. 5). A commercial BCG bed based on this development is now produced by the Royal Medical Corporation of Huntsville, Alabama.

C. Accelerometers and Calibration

Accelerometers for BCG must be capable of faithful sensing of motion of 1–5 cm/s^2 maximum in the typical heartbeat. This signal is super-imposed on zero mean level in the horizontal BCG, and on minus 1 g for vertical recording. Adequate rendition of the patterns requires that the accelerometer must be able to resolve at least one twentieth of the maximum, or .05 cm/s^2 (about 50 micro-g). Ideally, it should have unidirectional sensitivity in the

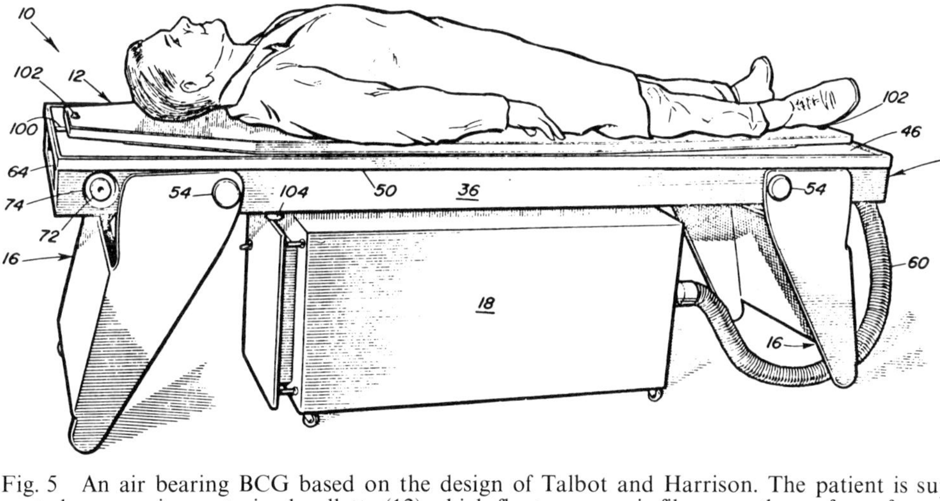

Fig. 5 An air bearing BCG based on the design of Talbot and Harrison. The patient is supported on a spring restrained pallette (12) which floats on an air film over the surface of an air bearing (36). Air for flotation is supplied by an industrial vacuum cleaner in the sound proofed enclosure (18), through a flexible hose (60). An accelerometer for sensing motion in the head-foot direction of the patient is mounted in the pallette. Coupling between the patient and pallette is improved by addition of either a footboard or foam plastic padding which also increases comfort of the subject. The legs (16) of this apparatus were removed so that it could be used on a flouroscope table for studies during cardiac catheterization. See Harrison et al (1969).

frequency range of 0–50Hz, since motion in directions orthogonal to the longitudinal axis of humans is of the same order of magnitude as along it. Direct current (DC) response is of great value for tilt calibration of the accelerometer using the gravitational field as a convenient primary reference. Sensitivity to motions with frequency components higher than 50 Hz is not justifiable in light of present knowledge of aortic flow patterns and heart movements. Intrinsic electrical noise must be sufficiently low that expected output signal due to BCG will be presented with adequate signal to noise ratio. This figure will vary depending on many factors, but 20:1 is a realistic lower limit for graphic recording. Finally, the accelerometer must be small and light in weight. It adds to the mass of the patient support structure, which for reasons already given must be kept to a minimum.

The accelerometer is an important part of any system for taking BCG and its selection deserves careful attention. There are a number of suitable commercial as well as home made units, and improved versions are introduced from time to time. Any purchaser of a commercial BCG bed should assure himself that the accelerometer employed by the manufacturers, as well as other features, meet standards necessary for his expected use. Rather than make recommendations of specific models of accelerometer, we believe it is more appropriate to indicate general classes of instrument which are and are not likely to be suitable. With this information in mind, one can then locate compilations of products of many different manufacturers to identify models which qualify for the application. One example of such compilation is published by the journal *Instruments and Control Systems*, Rimbach Publications, 56th and Chestnut Streets, Philadelphia, Pennsylvania 19139.

Accelerometers all contain a suspended seismic mass coupled to a surrounding case which is mounted on the member whose acceleration is to be measured. The electrical output signal proportional to acceleration is derived from forces developed between the seismic mass and the case as a result of movement of the latter. The methods by which forces produce the electrical signal provide a way of classification of accelerometers. The six most common classes are:

a. servo force balance
b. resistive
c. variable impedance or reluctance
d. differential transformer
e. piezoelectric
f. potentiometric

We have had favorable experience with accelerometers in the first two categories which met requirements just described. These units were manufactured by Donner, Endevco, Kistler, and Statham in the United States. There are doubtless similar accelerometers produced by manufacturers in other countries. Accelerometers in the other four classes are unlikely to meet requirements for use in recording BCG although there may be a few exceptions.

An extremely valuable property of the servo force balance and resistive accelerometers is their DC response. This characteristic allows us to take advantage of the earth's gravitational field for calibration of BCG records. The procedure is easily accomplished by tilting the patient support platform of the BCG in a careful, accurate way, then noting the resulting response of the graphic recorder.

Let θ denote the tilt angle in radians of the sensitive axis of the accelerometer from horizontal, and g the gravitational acceleration (980 cm/s^2). The acceleration signal generated by the tilt is then $y = g \tan \theta$.

Fig. 6 illustrates use of this principle for calibration of the accelerometer of a BCG bed. Advantage is taken of the relatively long length 1 of the bed platform, and accurately measurable blocks such as drills to raise its ends h for acceleration signals $= gh/1$. As a practical matter, several different values of h should be used in the procedure. The calibration factor for the recordings may then be determined from the slope of a least squares straight line fit to measured data points. If a secondary standard acceleration is available, such as a rotating unbalanced mass, its value may be determined after the tilt calibration. The rotating mass may then be used when more frequent, routine calibration checks are desired.

Sellers of commercial BCG beds should be required by the purchaser to furnish evidence of adequate calibration of the accelerometer. Graphs of the tilt calibration procedure as described above are acceptable documentation for a secondary calibration reference. If biomedical or clinical engineering consultation on accelerometer claibration is available, this may be helpful in judging adequacy of these or other procedures which may be employed. Accurate calibration of accelerometers for use in re-

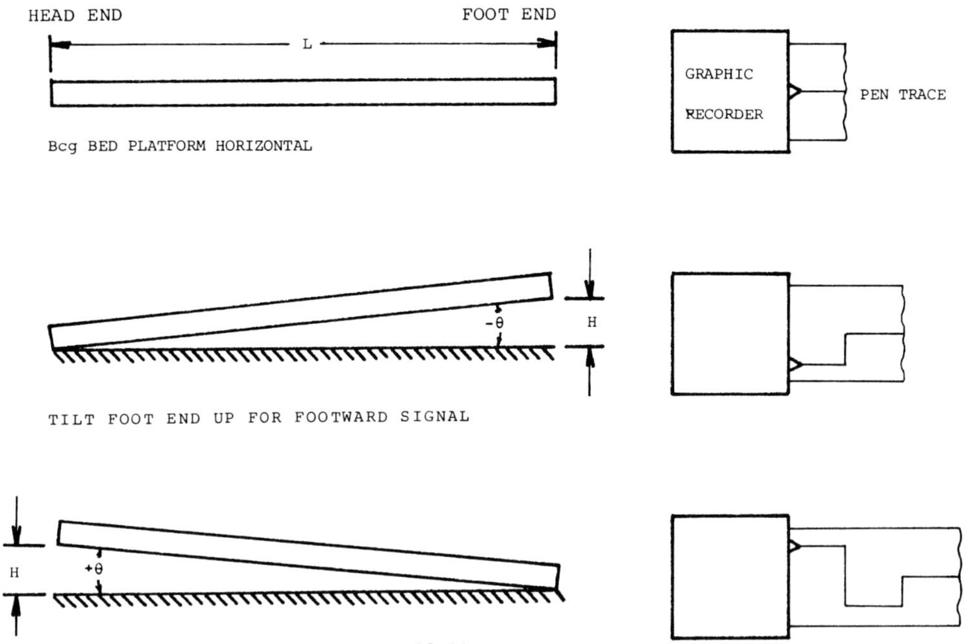

Fig. 6 Schematic diagram illustrating tilt procedure which takes advantage of the relatively long length L of the BCG platform and small, accurate blocks or rods H for accurate calibration of BCG. For a typical L = 200 cm, and two equal blocks H = 0.5 cm, five calibration points representing −4.9, −2.45, 0, 2.45, +4.9 cm/s^2 can be quickly obtained. The least squares linear fit to these points and their corresponding chart deflections will give an accurate estimate of calibration in cm/s^2 per cm of chart ordinate (see text). In practice, ambient vibration makes it necessary to estimate the midpoints of recorder pen traces or heavily filter signals.

cording BCG is a very important requirement, and replicate or periodic checks are completely justifiable. The procedure is analogous to the familiar millivolt calibration widely employed in ECG recording.

D. BCG in Noninvasive Cardiology

Capability for recording resting ECG is now quite common in medical offices, clinics, and all hospitals. Expanded facilities for ECG stress testing are also being set up with increasing frequency. In addition, growing awareness of physicians of the value of noninvasive techniques for cardiovascular evaluation seems likely to increase future demand for an availability of these services. Craige (1971) has indicated that the various techniques are best applied together in appropriate combination, not singly. It seems likely, therefore, that they will be offered by laboratories well equipped for multiple measurements in addition to ECG. At such locations there will be multi channel capability in a variety of signal processors, transducers, oscilloscopes, and recorders. BCG can be easily added to such laboratories, and in this environment it can make its most valuable clinical contribution. As Weissler (1974) points out, this technique provides the clinician with information on aortic blood flow dynamics which is not available from the other methods.

By virtue of its similarity to a bed or examining table, the BCG can be conveniently included in almost any noninvasive cardiovascular laboratory. If the existing couch or examining table serves merely for holding the patient during resting ECG recording, it may simply be replaced by the BCG. ECG and pulse recording can be done on it equally as well as on the other kinds of support where floor space is at a premium. For practical reasons, a BCG without long suspension wires or tall superstructure should be used. Air bearing or inverted pendulum designs previously discussed meet this requirement. The room itself should be located in a vibration free area if possible. If heavy ventilation or air conditioning machinery, elevators or escalators are nearby, some vibrations from their operation may show up in the BCG recording. Smith's comments (1974) on practical aspects of setting up BCG should be reviewed before making final plans. If the apparatus is obtained from a commercial source, its acceptance must be contingent on demonstration of satisfactory performance in the user's laboratory. If the purchaser is familiar with his own BCG pattern, or that of an associate who may be tested, the decision on acceptability is easily made.

Fig. 7 indicates schematically the principal elements required to record BCG in a laboratory for noninvasive cardiovascular testing. Another cardiac variable such as ECG, CP, or phonocardiogram is recorded simultaneously to provide a timing reference in cases where the BCG pattern is very abnormal. Each item of equipment must be thoughtfully located so that its controls are conveniently accessable to the operator who conducts the tests. This usually means that they must be close by the BCG itself since it is sometimes advisable for the operator to stabilize manually the movable patient support prior to recording. Such timely intervention will eliminate any large excursions which may be caused by the patient's unintentional movement. Viscous damping of the support is also valuable for this purpose if it is less than about 0.3 times the critical value (See Cunningham 1970).

Amplifiers for the BCG must be suitable for the accelerometer employed, and vary greatly according to the type used. Since the accelerometers are not self energizing, there must also be a source of power available for them. Selecting components for the BCG channel is not difficult for the user who has adequate electronic background. On the other hand, sellers of the amplifiers, recorder, or BCG can often make helpful recommendations. If services of a clinical engineer or medical physicist are available, their advice will also be of significant value.

Testing procedures for the resting BCG are not complex. Its place in the sequence of tests to be made on the patient depends on the protocol which has been selected for him. As a general rule, it is important that the patient rest in supine position for 5–10 minutes prior to recording so that cardiovascular transients

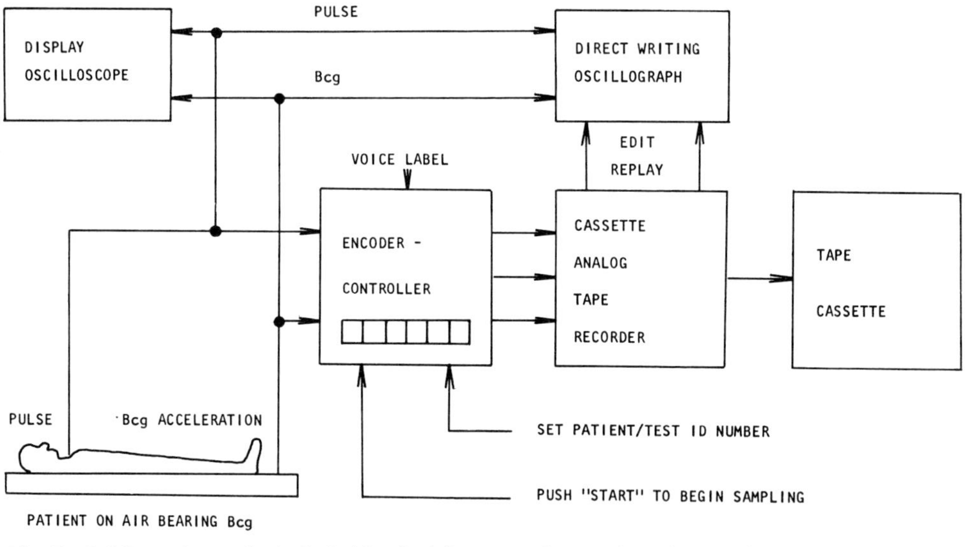

Fig. 7 BCG can be easily included in the laboratory for noninvasive cardiology. In most cases, there is already a display oscilloscope and multi channel direct writing recorder. If the BCG is like the airbearing design illustrated in Fig. 5, it can replace an existing examining table so that no increase in floor space is required. Additional elements for tape recording are employed in a procedure for computer analyses of BCG which is described in section IV.B.

due to position change have died out and a steady state has been reached. The physician or test operator must be skillful to detect and allay any apprehension which the patient may have concerning the test. It is easy to reassure the patient about this, since no demands are made on him other than he become as relaxed as possible while he lies on the BCG. If a standard resting ECG is to be made, it is advantageous to do this procedure first since the patient will stabilize for the BCG during the time required for this test. Of course, stress testing of any kind should be done only after the resting BCG has been recorded. Patients should be advised to abstain from eating and smoking cigarettes for at least two hours prior to the test, since these activities have immediate cardiovascular effects which cause a response in the BCG.

The resting BCG is ordinarily recorded with the patient performing two standardized respiratory maneuvers: eupnea and apnea, in that order. It is important to request the patient to breathe quietly, perhaps more slowly than usual if there is a tendency toward tachypnea. Two or three complete respiratory cycles should be recorded, for a total duration of 10–20 seconds. The operator should then immediately review the recording to check for any undesired artefactual accelerations due to involuntary movements of the patient. If these are present, a repeat recording should be made until the desired results are obtained.

A trial or practice run in which the patient is coached to hold his breath in the required respiratory phase is often helpful before recording the BCG in apnea. Successful recordings in apnea can usually be secured by means of this procedure even in patients who are very diffident. Rehearsal also allows the operator to observe the length of time which the patient is capable of comfortable breath holding. The recording can then be precisely timed by stopwatch to last no longer. Breath may be held in full inspiration, expiration, or mid expiration as desired. Care is sometimes required to prevent the patient from performing a valsalva maneuver during full inspiratory apnea. Again, each recording should be reviewed immediately and repeated if necessary to obtain satisfactory results. Charts should be carefully marked with patient identification and respiratory maneuver during the review. An experienced operator can accomplish the BCG test, not including the initial rest period, in approximately 5 minutes with most patients.

IV. Interpretation of BCG

A. Subjective Classification of BCG Based on Contour

A person's BCG pattern is as individualized as his signature. Somewhat paradoxically, the similarity of most BCG from young subjects in good cardiovascular health is also very striking. Both individually and as a group, these recordings from healthy people differ greatly from those of patients with many kinds of heart disease. A trained observer can do a good job of picking out such patients merely by inspection of their BCG recordings.[4] This is by no means true of all cardiac patients in general. The author (1974) found that well qualified reviewers were unable on the average to distinguish a heterogeneous group of patients who had a variety of cardiac disease from an age matched group of healthy subjects by means of their BCG pattern alone.

Any algorithm for subjective classification of BCG as normal, abnormal, or categorical variations of these, requires judgements about each individual heartbeat pattern. Fig. 2 is typical of patterns from young, healthy, hence "normal" subjects. Experience of each reviewer influences the extent to which variation in proportion of the waves may be acceptable within limits of his personal rubric for "normal." For example, a more pronounced $F_2 - H_o$ feature than that illustrated in the figure may be accepted by most reviewers as within normal limits. Or, some minor degree of splitting or doubling of the J peak in mid systole may be permissible. However, these and other variations are often strongly expressed in records from cardiac patients. The reviewer must therefore exercise considerable care in deciding just how heavily to weight each effect.

Fig. 8, from Starr and Noordergraaf (1967), illustrates various abnormalities of BCG waves about which many experienced reviewers would be likely to agree. This subjective identification of abnormalities has been found to be surprisingly repeatable for experienced reviewers who had worked closely together (Sherwin et

4 For example, BCG patterns from cardiac patients with advanced, symptomatic arteriosclerotic cardiovascular disease are almost invariably abnormal. On the other hand, those from young subjects with minimal aortic insufficiency are likely to be normal.

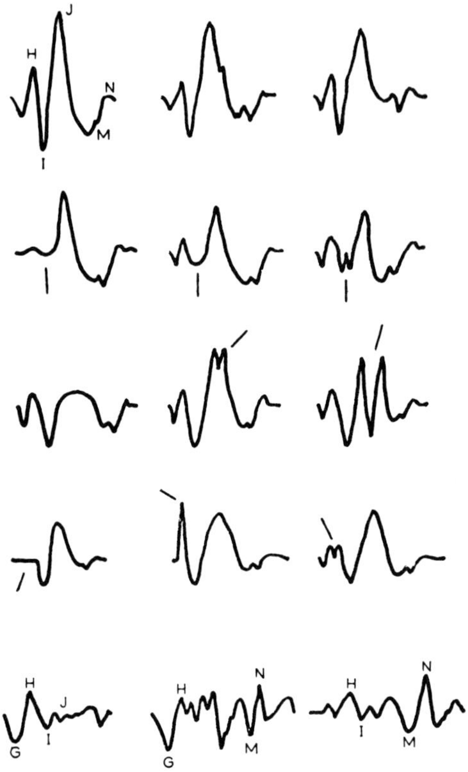

Fig. 8 Normal and common abnormal forms of ULF force BCGs drawn to scale from actual records. The lines point to features of interest. All have been aligned by means of simultaneous ECGs. Top row: Normal records. Note the slight notching of some records, its positions may vary. Second row: I wave abnormalities, from left to right, the I wave is absent, shallow and rounded, notched. Third row: J wave abnormalities: low rounded, notched and divided. Fourth row: H wave abnormalities: absent, exaggerated, and notched. Fifth row: confused records with multiple abnormalities. In such records the individual waves cannot always be identified with confidence, but the preferred interpretation is indicated by letters. From: *Ballistocardiography In Cardiovascular Research*, by Isaac Starr and Abraham Noordergraaf, J. B. Lippincott Company, Philadelphia/Montreal, 1967. (With permission of authors and publisher.)

al 1969). The repeatability has also held for those who are less closely associated (Harrison et al 1974), but for these there was a much higher incidence of disagreement on exact limits of normality. The difficulty of arriving at explicit standards of pattern abnormality which can be accurately and uniformly employed in subjective rating of BCG is a real disadvantage of the technique.

The problem of subjective characterization of abnormality in single heartbeat patterns is in the final analysis solved by each reviewer in his own way. Once the determination has been made, results from it can readily be employed in an algorithm for rating as a whole the collection of heartbeats from the patient. The method described by Starr (1967) is a good example of various approaches which have been used. Starr's procedure assumes that respiratory cycling can be determined either directly from the BCG or from a simultaneously recorded pneumogram. Workers agree that abnormalities are more likely to be found in the smaller patterns recorded during expiration than in the larger size complexes usually found during inspiration. Because of this fact, care must be exercised to include complete respiratory cycles in the evaluation to avoid bias in application of the procedure.

When all complexes in the respiratory cycle are normal, the BCG is rated grade 1. When a minority of the heartbeats are abnormal, the rating is 2; when the majority are abnormal, 3. If the heartbeat patterns are so grossly abnormal that the major waves of the BCG cannot be recognized without aid from some accompanying recording such as pulse or ECG, then the rating is grade 4. This method has been employed in a number of studies which showed an increasing incidence of the abnormal ratings among asymptomatic older persons, and a statistically significant prognostic power of the abnormal ratings for future development of coronary heart disease. Such results have thus linked subtle deficits in cardiac function to accelerated cardiovascular aging and later development of coronary disease. The subjective methods for rating BCG have therefore been useful and effective in spite of their clearly evident limitations.

B. Computer Interpretation of BCG

Disadvantages of categorical subjective evaluation of BCG waveform and laborious manual measurement of its waves were recognized by some of the early workers. They were also aware of uncertainties about its relationship to causative hemodynamic events and the likelihood that methods then in use caused serious recording artifacts. Available effort was understandably directed toward correction of these known limitations instead of toward exploration of new methods, such as those involving computers, for extracting information from the BCG. An equally good reason for inaction in this pursuit was the general unavailability of computerized methodology for accomplishment of this goal, and the unfamiliarity of most biomedical researchers with that which was available.

While the first efforts at computerized analysis of electrocardiograms began in the late 1950's and early 1960's, even they were confined to a few laboratories specially equipped for this task. It was not until later in the 1960's that necessary electronic hardware became widely enough available to stimulate application to a variety of biomedical research recording. The first of such equipment to be applied to research on the BCG was the computer of average transients (CAT). Jackson (1967) used a CAT to obtain average BCG patterns of healthy young men in the immediate post exercise recovery period. Movement artifacts in the acceleration recording are often caused by heavy breathing just after strenuous exercise. In Jackson's series, these random fluctuations were efficiently eliminated by the CAT when triggering was controlled by a simultaneously recorded ECG. Jackson also demonstrated that no more than 13 heartbeats of BCG recordings of healthy young men were required to generate a stable average pattern either at rest or after strenuous exercise.

Another application of average transient computing to BCG was reported by Rautaharju (1966). His article illustrates separation of the "cardiac component" from the mixed respiratory and cardiac displacement BCG recorded during ordinary breathing. When the ECG was used as the trigger signal, the CAT averaged out respiratory motions which were many times larger than the desired cardiac signal. When a respiratory signal was employed to trigger the CAT, cardiac motions were removed and a kind of pneumogram remained. Use of the CAT, which is actually a small special purpose digital computer, gave the first satisfactory rendition of a ulf BCG displacement recording during breathing. Previous attempts to separate cardiac and respiratory components in ulf displacement BCG by analog means have not been successful.

Johnson et al (1969) applied average transient computing in a small general purpose digital computer to study BCG of 13 patients who were successfully treated for congestive heart failure. Prior to treatment, heartbeat patterns were abnormal, highly variable and difficult to interpret. Computer averaging of 10–20 heartbeats, using ECG QRS triggering, produced a stable, more easily interpretable BCG. After treatment for heart failure, the average heartbeat patterns of the patients more closely resembled normal ones obtained from healthy young subjects.

Use of the CAT or other average transient technique demonstrated advantages of computer methodology for extracting valid cardiovascular data from BCG which could be heavily contaminated with undesired noise. These investigations also demonstrated that stable average patterns could be obtained with samples of 10–20 heartbeats. Such results were important for later research which explored inter-heartbeat variability and other properties of BCG. With digitization rates of even 1000/s, total number of data points for averaging, feature recognition, and estimating variability was within the practical working capacity of many general purpose computer systems.

The author (1972) made use of a small, general purpose computer (Varian 620, 16K memory) for development of a BCG record reading program. Eight to thirteen consecutive heartbeats were digitized by the computer operator, who compared their images on a core display oscilloscope with desired selections from an analog magnetic tape recording. A linc digital magnetic tape was produced by this step. This tape, which contained heartbeat samples from both eupnea and apnea of the patient, served as the input vehicle for data for the record reading program.

The BCG analysis program ran in the same computer. It detected an epoch for each heartbeat at the onset of the accompanying percutaneous carotid arterial pulse pattern. From this epoch, searches were made by the computer for H, I, J, M, and N waves of the BCG. After recognition of these features in each heartbeat, acceleration amplitudes of the various waves were calculated, and timing of the wave tips determined in relation to the carotid upstroke. Additional measurements of jerk, cross correlation coefficients with a reference pattern, and inter heartbeat variability were also determined. An average BCG pattern was then calculated, and similar analyses made of it as for the 8–13 individual heartbeats in each sample. Several hundred BCG from normal subjects and a variety of cardiac patients were used to develop effective computer algorithms to recognize and measure desired variables of the pattern.

A procedure for computer processing of the ulf BCG which employs timing information from both the Frank orthogonal ECG and the percutaneous carotid arterial pressure pulse has been described by Bancroft et al (1973). Their procedure involves a human operator to control several aspects of the process such as archiving of patient identification numbers and quality control of feature recognition of ECG, BCG and pulse. Three epochs of each cardiac cycle are located by the program operating in Xerox Sigma 7 computer for use in searches for the BCG wave tips. They are the Q wave of the ECG, the carotid upstroke, and carotid incisura. Location of each of these points as determined by the computer is indicated on an oscilloscope for review by the operator, who may accept, reject, or correct them. A digital tape entry containing identification, epoch locations, and slightly over 3 seconds of BCG data is then made by the Sigma computer. Studies on the digital tape are later analysed in batch mode by a program which locates BCG wave tips and reports amptitudes, slopes, and timing for the heartbeats sampled.

A further development (Harrison et al 1974) of the previously described (Harrison et al 1972) computer program for analysis of BCG and carotid pulse has been made by the author. This approach dispenses with the operator–computer interaction previously used for sampling specific heartbeats to be analysed. In its place, a specially developed electronic unit called an encoder-controller was substituted to control the analog tape recording of BCG and pulse patterns. Fig. 7 illustrates the position of the encoder-controller in signal pathways of the patient testing laboratory. The encoder-controller produces a rigidly formatted analog tape recording, which, after digital conversion, simplifies computer editing steps to obtain the identification number and desired data points. Fig. 9 shows a reproduction of the BCG and pulse data sampled under control of the encoder, and the 7 digit BCD identification code it produces.

Analog to digital conversions of the tapes

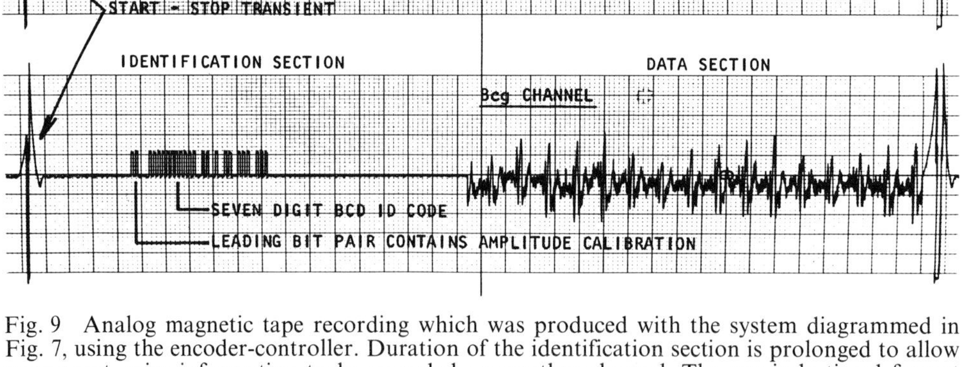

Fig. 9 Analog magnetic tape recording which was produced with the system diagrammed in Fig. 7, using the encoder-controller. Duration of the identification section is prolonged to allow concurrent voice information to be recorded on another channel. The precisely timed format produced by the encoder-controller simplifies the subsequent computer editing step which works from a digital translation of the above recordings.

formatted by the encoder-controller are input vehicles for an editing program which operates in batch mode in an IBM 370/145 computer system. The edit program decodes the ID numbers and recognizes beginning and end of BCG and pulse data for transfer to another industry compatible digitial magnetic tape. This tape contains a readily accessable "directory" of ID numbers of patient data archived on it, so that any particular study may be selectively analysed by the BCG record reading program. The directory tape is dimensioned to contain up to 200 of the studies illustrated in Fig. 9, and may be successively updated until this number is reached.

Analyses of studies archived on the directory tape are made by a revised version of the record reading program previsouly described (Harrison et al 1972). This program operates in an IBM 370/145 computer system. Particular studies on the directory tape which are to be analysed are identified by means of punched card input to the program. These cards also contain necessary data such as calibration factors and title headings for the analysis report produced by line printer. Table 2 illustrates a typical report, which summarizes detailed, individual analyses of heartbeats which were sampled under encoder control. This analytical program has been intensively tested on a full spectrum of BCG and pulse patterns ranging from normal to extremely abnormal. Algo-

Table 2 This is an example of a report on a computer analysed BCG. It is based on 16 heartbeats, sampled by means of the system described in section IV.B. utilizing an encoder-controller unit. Summary calculations are based both on the computer averaged pattern formed from the 16 heartbeats and on average values measured in each separate beat. Blood pressure and values of maximum derivative of the percutaneous carotid arterial pressure pulse (MPD) are based on the blood pressure of the subject obtained immediately prior to the test. Information derived from time-domain features of BCG such as time intervals, magnitudes, contour, and interheartbeat variability is included.

```
                COMPUTER READINGS OF BALLISTOCARDIOGRAMS   V.04.25.66

NO. *2301411          MINNEAPOLIS PHYSICAL CONDITIONING STUDY
ANALYSIS DATE  7 OCT 1974

REMARKS-
  CDP    23014                SEGMENT  1   TEST     1

16 HEARTBEATS, AVERAGE DURATION 1.018, PULSE RISE TIME 0.085 SECONDS

HEARTBEAT DURATIONS IN SEQUENCE, SECONDS
 1.045 0.960 0.970 1.065 1.090 1.035 0.970 1.045 1.065 1.015 0.910
 1.000 1.070 1.040 0.945 1.060

VALUES CALCULATED USING AVERAGED HEARTBEAT

MEAN BLOOD PRESSURE    ABSOLUTE VALUE ACCELERATION    PULSE RISE TIME
     80. MM HG                0.6022 CM/SEC2                0.085

CROSS CORRELATIONS-
 RF=-0.520  RG= 0.856  RH= 0.790  ROH= 0.472    MPD= 1099. MM HG/SEC

HI MAX JERK =  83. CM/SEC3 AT  0.020 SEC.

TIME IN SECONDS-
CUY = 0.005    CUH =-0.025    CUI = 0.035
CUJ = 0.105    CUM = 0.340    CUN = 0.410

AMPLITUDE IN CM/SEC2-
XH = 0.9       XI =-2.1       XJ = 2.9       XM =-1.8       XN = 0.9
HI = 3.0       IJ = 5.0       JM = 4.7       MN = 2.7

AVERAGE VALUES CALCULATED FOR THE HEARTBEATS

CROSS CORRELATIONS-
 RF=-0.375  RG= 0.797  RH= 0.692  ROH= 0.427    MPD= 1143. MM HG/SEC

HI MAX JERK = 106. CM/SEC3 AT  0.018 SEC.

TIME IN SECONDS-
CUY = 0.004    CUH =-0.020    CUI = 0.037
CUJ = 0.107    CUM = 0.337    CUN = 0.426

AMPLITUDE IN CM/SEC2-
XH = 1.1       XI =-2.3       XJ = 3.1       XM =-1.9       XN = 1.1
HI = 3.4       IJ = 5.4       JM = 5.0       MN = 3.0

STANDARD DEVIATIONS OF CROSS CORRELATIONS AS FRACTION MEAN VALUES

 SF= 0.3356   SG= 0.1284   SH= 0.1717   SOH= 0.1200
```

rithms for feature recognition and other operations have been refined in this testing so that valid, quantitative data may be obtained from very abnormal patterns which present significant difficulty for subjective reviews or manual measurements. The program is currently being used to analyse records obtained in epidemiological studies of coronary heart disease, and in other research projects.

'Glossary/Terminology'

1. ballistocardiogram – a periodic "whole body" motion pattern produced by action of cardiovascular forces. The motion may be displacement, velocity, or acceleration of the structure supporting the patient, and is usually recorded in the direction of the long axis of the body.
2. ballistocardiograph – an apparatus for recording the ballistocardiogram. A ballistocardiograph is essentially a very light weight bed-like platform on which the patient lies. The platform is supported in a way which minimizes its coupling to the earth, at least in the direction in which motion is to be recorded.
3. BCG – abbreviation for either ballistocardiogram or ballistocardiograph.
4. transcutaneous arterial blood pressure pulse – localized infrasonic motion or force pattern sensed at the surface of the body in the vicinity of an artery by an appropriate transducer, and recorded on devices such as oscilloscopes, strip chart recorders, or magnetic tape recorders.
5. acceleration – second time derivative of displacement. If $y\ (t)$ is displacement of a body, its acceleration is $\mathrm{d}^2y/\mathrm{d}t^2$.
6. jerk – third time derivative of displacement, derivative of acceleration. If $y\ (t)$ is displacement, jerk is $\mathrm{d}^3y/\mathrm{d}t^3$.
7. damped fundamental natural frequency – the frequency of decaying vibration of any material structure (such as the human body) which results when a steady load is suddenly removed or applied.
8. accelerometer – a transducer for measurement of acceleration. Usually a small, lightweight case containing a suspended seismic mass with one degree of freedom. An electrical signal for output is derived from relative force or motion between the seismic mass and case.
9. computer of average transients – a device which produces ensemble average patterns of periodic input waveforms presented to it. Patterns are aligned in reference to a synchronizing epoch so that features which are random in relation to the epoch are averaged to zero value in the ensemble average pattern.
10. eupnea – free, easy respiration of the kind found in normal, healthy resting subjects.
11. apnea – absence of respiration, such as during intentional cessation of breathing.

References

Abramson E: Die Rückstosskurve des Herzens. Scand Arch Psysiol 66, 191–224 (1933)

Angenheister G, Lau E: Seismographische Aufnahmen der Herztätigkeit. Naturwissenschaften 16, 513–515 (1928)

Baan J, Manchester J H, Shelburne J C: A quantitative relationship between the H-I slope of the head-foot Bcg and the initial acceleration of flow in man. Bibl Cardiol 29, 55–69 (1972)

Baker, Benjamin M, Scarborough, William R, Davis, Frank W Jr, Mason, Robert E, Singewald, Martin L, Walker, Strother H, Murphy, Edmond A, Harrison, Walter K Jr, Sherwin, Roger W: Ballistocardiography and ischaemic heart disease: predictive considerations and statistical evaluation. Proc R Soc Med 60, 1290–1297 (1967)

Bancroft W H Jr, Tucker, Marjorie, Jackson D H, Eddleman E E Jr: Automatic computer processing of ultralow frequency ballistocardiograms. Bibl Cardiol 32, 1–10 (1973)

Burger H C, Noordergraaf A: Physical basis of the low-frequency ballistocardiograph. Am Heart J 46, 71–83 (1953)

Craige, Ernest: Clinical value of apex cardiography (editorial). Am J Cardiol 28, 118–121 (1971)

Cunningham D M: Specifications for the ultra-low frequency ballistocardiograph. Bibl Cardiol 26, 343–353 (1970)

Cunningham D M, Smiley P C: A hydrostatic airbearing ballistocardiograph. J Appl Physiol 16, 756–758 (1961)

Dock W, Taubman F: Some Technics for recording the ballistocardiogram directly from the body. Am J Med 7, 751–755 (1949)

Douglas C G, Haldane J S, Henderson Y, Schneider E C: Physiological observations made on Pike's Peak, Colorado, with special reference to adaptation to low barometric pressures. Trans R Soc (London) Series B, 302, 185–309 (1913)

Freberg C R, Kemler E N: Elements of Mechanical Vibrations. Wiley, New York 1949

Adv. cardiovasc. Phys., vol. 5 (Part III), pp. 59–83 (Karger, Basel 1983)

Mechanoventriculography: Theory, Evaluation and Significance of the Regional Distributions of Myocardial Constitutive Properties and Blood Pressure in the Left Ventricular Chamber[1]

D. N. Ghista, G. Ray, H. Sandler
Houghton, Michigan; Milwaukee, Wisc.; Moffett Field, Calif., USA

Contents

[1] The support of NASA grant no.: NSG2041 for this research project is acknowledged.

Abstract

In order to compute the distribution of the stress (σ) dependent myocardial modulus (E) of the left ventricle, the instantaneous inner wall deformations of an instantaneous (solid) finite element model of the LV chamber are matched with the known instantaneous deformations derived from the sequential cineangiographic (or echocardiographic) outlines of the LV, to obtain the instantaneous regional distructions of E based on the values of the constitutive coefficients a and b of the constitutive relation $E = a + b\sigma$, it is possible to distinguish pathological (ischemic-infarcted) myocardial elements from normal elements. In order to develop a noninvasive technique, not requiring monitoring of the chamber pressure, it is shown that the value of the coefficient b is independent of the chamber pressure, while the value of the coefficient a can be determined (without knowledge of the chamber pressure) with adequate sensitivity to help distinguish between normal and pathological elements.

The computation of the relative blood pressure distribution in the LV chamber is carried out by means of an instantaneous (fluid) finite element model of the LV chamber; the model's fluid velocities at the endocardial boundary are computed from the sequential LV outlines obtained by cineangiography or echocardiography. The computation of the relative pressure (normalized with respect to the pressure at the aortic valve interface) does not require knowledge of the chamber pressure; the method is hence noninvasive. The distribution of the normalized chamber pressure provides a graphic illustration of the outcome of the contractile action of the various myocardial segments of the LV, in developing a pressure distribution conducive to effective emptying of the LV chamber. To this end, a set of parameters based on the intraventricular pressure distribution is derived to provide a new intrinsic index of LV contractility.

I. Need and Utility

Left ventricular (LV) mechanical performance is generally characterized by a myriad of extrinsic indices, such as cardiac output, ejection fraction and pressure-volume relationships. However, in order to provide the etiological basis of LV performance, there is a need for developing effective sensitive intrinsic diagnostic mechanical indices that can differentiate between normal and diseased myocardium.

The LV mechanical indices which we propose consist of (1) regional distribution of the mechanical constitutive properties of the passive myocardium during diastole and (2) the distribution of the instantaneous pressure of blood in the LV chamber which will be indicative of the LV pumping efficiency during systole. We will also demonstrate that these indices will provide sensitive assessment of the LV pathology, and thus help provide guidelines for the remedial measures.

These two intrinsic indices would be useful clinically, (1) to delineate the location and extent of the pathological myocardial segments that have been affected by coronary artery disease, and (2) to provide an assessment of LV pumping efficiency, afford a basis for coronary by-pass surgery and to evaluate its efficacy. Additionally, both the indices will provide a quantitative basis for characterizing cardiac conditioning and deconditioning. Furthermore, if the formulation of these indices indicate that they could be derived by noninvasive monitoring procedures (such as two dimensional echocardiographic imaging), then these would constitute sensitive, quantitative and noninvasive cardiac diagnostic methods.

II. The State-of-the-Art Versus the Advances Afforded by our Current Investigations

A. Diastolic (Solid Mechanics) Analysis of the Left Ventricle to Determine Its Passive Elastic Stiffness

A traditionally employed LV index is its compliance or distensibility expressed by the ratio of change in volume of the LV chamber to its change in pressure ($\Delta V/\Delta p$). A decrease in the value of this index has been associated with coronary artery disease and myocardial infarction (see Diamond and Forrester 1972).

Usually the LV compliance is represented by a single value of the ratio of the volume-to-pressure change, for the entire diastolic period.

However, even a cursory observation of the successive frames of a cineangiogram and the associated catheter obtained chamber pressure changes (exemplified by a plot of associated volume and pressure changes shown in Fig. 1) reveal that the LV compliance during the diastole varies continuously in a nonlinear fashion. Thus it has been noted by Mirsky and Parmley (1974) that the LV compliance becomes a nebulus term unless the pressure at which it is evaluated is stated.

It may, however, be recognized that the volume change of any distensible chamber, such as the human LV, under a given pressure load, should be the function of the elastic properties (or mechanical stiffness) of the chamber wall (i.e., the LV myocardium). Thus, a change in the order of magnitude of the diastolic compliance (resulting from the altered LV volume change responses to LV pressure change due to heart disease) is interpreted as due to altered elastic properties of the LV myocardium. Mirsky and Parmley (1974) have reported such changes in the overall elastic stiffness of the relaxed LV myocardium in various diseased situations. The analytical model, for the computation of this overall elastic stiffness, assumed either a spherical or ellipsoidal geometry of the LV chamber.

Ghista et al (1975a, b) have also developed two such models for the evaluation of the instantaneous overall stiffness of the LV myocardium. In one formulation (Ghista et al 1975a), the instantaneous (chamber pressure related) stiffness value of the LV myocardium was obtained by matching the instantaneous strain response of an ellipsoidal model with the in vivo strains in the chamber wall computed from the dynamic LV (cine derived) geometry. The instantaneous LV chamber pressure, required to determine the model strains, was obtained by catheterization. In the other study (Ghista et al 1975b), the instantaneous elastic stiffness was determined at the instants corresponding to the occurrence of the first and the fourth heart sounds. Therein, the primary frequency of a spherical LV model, whose geometrical parameters were obtained from cineangiography, was matched with the corresponding frequencies of the appropriate heart sounds. Various quasinormal and diseased LVs were categorized by means of a diagnostic matrix comprising of elastic stiffness, chamber size and the ratio of the wall thickness to chamber size.

Regional stress dependent moduli of the myocardial elements:

Inspite of the demonstrated physiological significance of these models, they do not

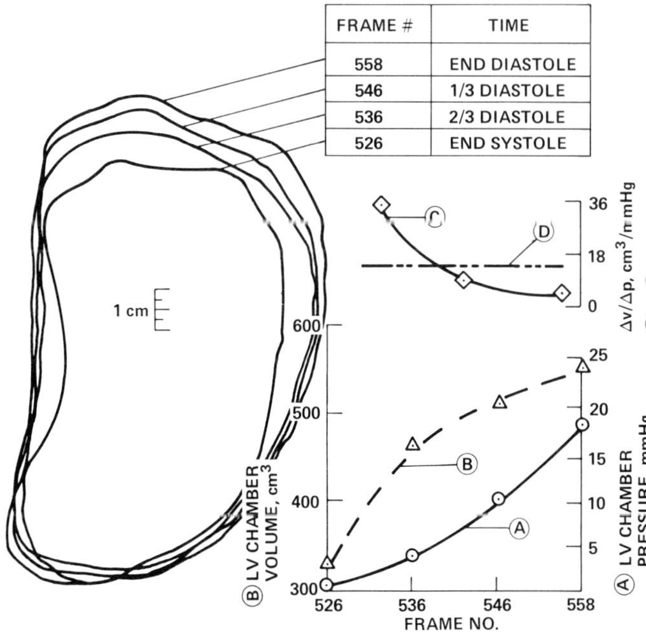

Fig. 1 Sequential traces of the left ventricle of a subject during diastole. A. LV chamber pressure variation during diastole. B. LV chamber volume variation during diastole. C. Variation of compliance $\Delta V/\Delta P$ during diastole. D. LV compliance based on the ratio of overall volume to pressure change at end diastole.

provide the regional variations of the myocardial elastic modulus versus stress relationship, which are essential for detecting the locations and extent of the diseased myocardial elements. Coronary artery disease and the resulting myocardial ischemia or infarctions would only change the elastic constitutive (stiffness versus stress) relationship of the affected regions, while leaving it unaltered for the rest of the myocardium. In fact, the evaluation of the regional distribution of the "elastic stiffness versus stress" relationship for the myocardium would not only help detect the extent and severity of the ischemic or infarcted myocardial zones, but would thereby also provide guidelines for evaluating the effectiveness of coronary by-pass surgery. Our model and associated analysis, that are presented in this chapter, cater to the above delineated need. Specifically, then, the methodology and its application for determining the modulus and principal stress, at each cineangiographic frame instant during diastole are presented. Therefrom will be developed the constitutive relationship, of the modulus versus principal stress, for each element.

B. Systolic (Fluid Mechanics) Analysis of the Left Ventricle to Determine the Blood Pressure Distribution Inside the Chamber —Its Relevance and Outline

Besides the analytical models of the LV outlined above, several three dimensional finite element models of the LV deformation have been developed by Ghista and Hamid (1977), and Ghista et al (1980). All these models have utilized cineangiography for the reconstruction of the three dimensional LV chamber outline. However, even these models only analyzed the passive (diastolic) responses of the myocardium. Even though the characteristics of the active myocardial contraction and the plausible constitutive relations of the myocardium during systole have been outlined by Ghista (1974), a comprehensive analytical model for the dynamic activation of the three dimensional LV chamber is lacking.

We have therefore, in this chapter, adopted an alternative approach to the analysis and extraction of biomechanical information regarding normal-pathogenic functional LV mechanics during systole, by performing a fluid mechanical analysis of the blood pressure distribution due to blood motion in the LV chamber. Such a fluid mechanical analysis provides a pumping efficiency index, in terms of a favorable intraventricular systolic pressure gradient. The basis of the hypothesis regarding the existence of this intraventricular pressure gradient is outlined below.

The existence of an intraventricular pressure gradient and its physiological significance has not hitherto been spoken of (and perhaps not acknowledged) in physiological parlance. The pressure in the left ventricular chamber (or for that matter in any cardiac chamber) is referred to and implied as being uniform. However, there is continuous motion of blood in the chamber throughout the cardiac cycle.

When either valve of the LV chamber is open, there is blood motion due to net increase or decrease of the chamber volume. However, even during the isovolumic (contraction and relaxation) stages, the chamber wall motion (caused by propagating electrical excitation waves) imparts considerable motion to the contained blood, which must result in a pressure gradient distribution in the chamber. It is the existence of this gradient, just prior to the opening of the aortic valve, and early systole that we will concern ourselves with in this chapter.

We will provide the analysis for the calculation of this pressure gradient and back it up with actual recordings in animals. The physiological significance of this gradient at the time of opening of the aortic valve and early systole, is that the existence of a favorable gradient toward the aortic valve opening will contribute to a higher pumping efficiency. On the other hand, the pumping efficiency will be decreased if this intraventricular pressure gradient is nonexistent or even reversed in direction. Therefore, it is reasonable to expect that during early systole, for a normal LV, the wall movement will be such that a favorable intraventricular pressure gradient will be generated. For a pathological LV, it may be hypothesized that the wall movement pattern will not be conducive to the generation of favorable pressure gradient.

In order to analytically investigate the above considerations, we (Ray et al, 1976b) have earlier performed a finite element analysis of the fluid potential distribution (hence the pressure and velocity distribution) in the LV chamber during a small time interval, just prior to the opening of the aortic valve. In this analysis, the fluid (blood) was assumed to be inviscid and incompressible. The computer analysis used the instantaneous kinematic boundary conditions of the LV wall velocity (derived from instant-to-instant wall motion obtained by single plane cineangiography) to compute the potential distribution.

Such an analysis was performed for the LV of a patient with myocardial infarctions, both before and after the coronary by-pass surgery. The results of the analysis showed that before surgery there was a large region in the chamber wherein the potential gradient was adverse and not conducive to ejection. However, the results of the analysis of the post operative angiographic outlines of the same patient showed a modest improvement of this gradient favoring ejection.

This computer model analysis strongly suggested that whereas in a normal LV there could be a favorable dynamic pressure gradient towards the aortic valve immediately prior to the valve opening, such a pressure gradient could be reduced or even reversed due to coronary artery disease.

A review of previous literature pertinent to the experimental investigation of the intraventricular pressure is given below. Gauer (1950) and Gauer and Henry (1964) reported that in a normal animal, the pressure recordings at the outflow region of the LV and the aortic root were congruent. However, a pressure drop of the order of 70 mmHg between the two recordings were observed when a circulatory shock was induced. Brock (1957) reported on intraventricular pressure gradients in patients with suspected aortic stenosis. However, subsequent postmortem examination revealed no obstruction in the outflow tract. White et al (1967) demonstrated the existence of high pressure gradients between the LV and the aortic root, with the administration of isoproternol in experimental animals. The existence of an intraventricular pressure gradient was also observed by them, without their attaching any importance to it, when they reported a fall in the systolic ventricular pressure to the arterial level with the gradual withdrawal of the retrograde catheter from the apex to the outflow tract. The main purpose of their study was, however, to report the pressure gradient between the LV chamber and the aortic root, rather than to investigate the significance of the intra-ventricular pressure gradient.

Preliminary experiments on seven mongrel dogs by Ray et al (1977) have consistently exhibited the following observations:

(1) There exists a positive pressure gradient (higher near apical region and lower near the aortic valvular region) in normal (control) animals. Fig. 2 shows such a typical pressure recording.

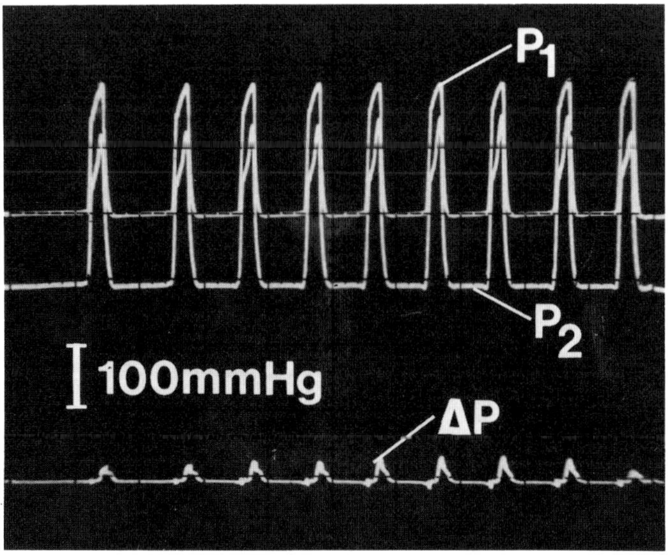

Fig. 2 Typical pressure recording in the dog left ventricular chamber showing the existence of an intraventricular pressure gradient. P_1, P_2 respectively represent catheter recorded pressures near the apex and the aortic valve respectively; $\Delta P = P_1 - P_2$. A positive pressure gradient (high near apex and low near AV) of as much as 15 to 20 mmHg is observed during peak ejection. This phenomenon has been observed in all the seven dogs.

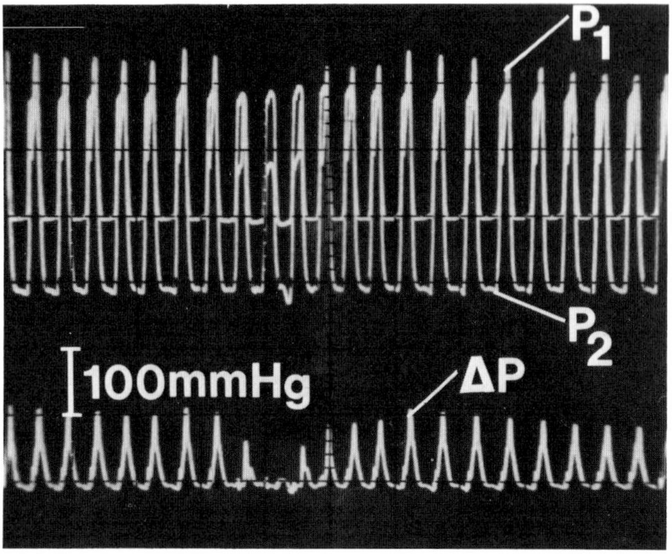

Fig. 3 Dramatic increase of intraventricular pressure gradient following the administration of isoproternol; this phenomenon was observed in all our experimental animals.

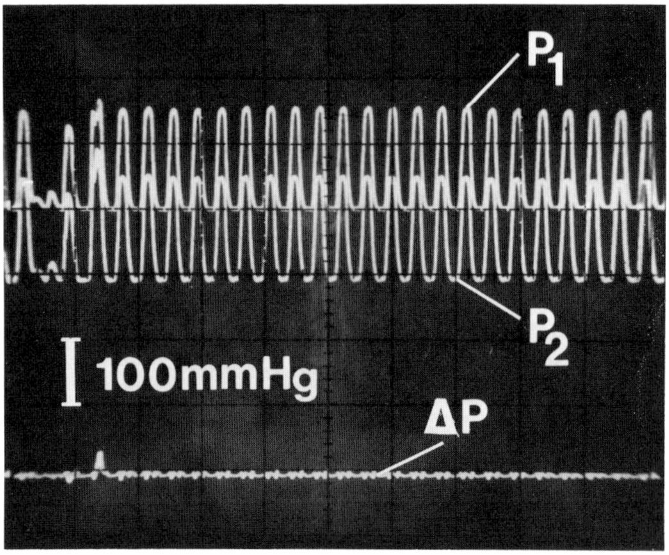

Fig. 4 Obliteration of the pressure gradient after ligation (for 28 min) of the left descending coronary artery; this phenonemon was also observed in all our experimental animals.

(2) With the administration of isoproternol, this positive pressure gradient is dramatically magnified, signifying vigorous pumping activity shown in Fig. 3.
(3) With the experimental production of the coronary artery occlusion (by ligation of the coronary artery), the intraventricular pressure gradient was invariably found to be obliterated (see Fig. 4). In some instances, this pressure gradient was actually found to reverse (see Fig. 5). A brief discussion of the methodology of the experimental procedure and the findings are provided in Appendix A.

In this chapter, we will present a more rigorous, two-dimensional, finite element model (relative to that of Ray et al 1976b) for the computation of the distribution of the intraventricular pressure gradient in the LV chamber during a small time interval at early systole just prior to the opening of the aortic

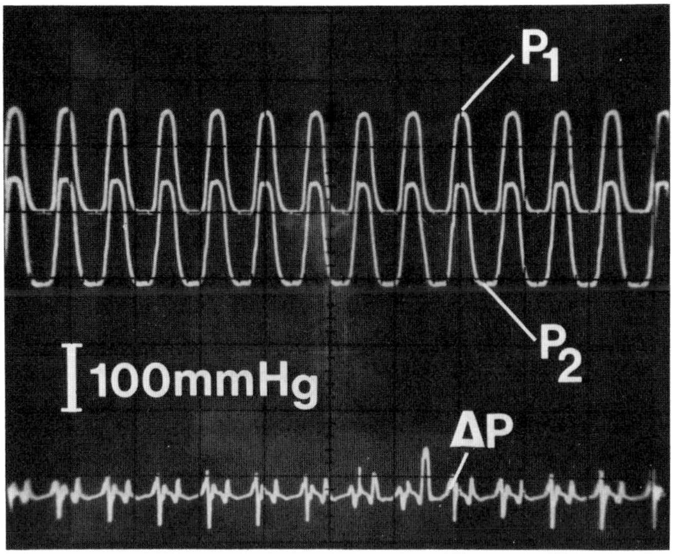

Fig. 5 Reversal of pressure gradient 45 min after ligation of left descending coronary artery (observed in 4 of the 7 dogs).

valve. The blood in the LV chamber is considered to be viscous and incompressible. During the small interval of time, between two successive frames, the blood flow is considered to be steady within the instantaneous chamber boundary. The instantaneous chamber wall velocity and hence fluid velocity at the boundary are computed from the monitored motion of the endocardial boundary at the two successive instants. In the governing Navier-Stokes equation, the convective terms are included in the computation.

III. Data Acquisition and Analysis Procedures for Determination of the Above Mentioned Indices

A. Solid Mechanical Analysis of the LV During Diastole to Determine the Myocardial In Vivo Constitutive Properties

The data needed for this analysis consist of the instantaneous left ventricular chamber sequential cineangiographic outlines during diastole, and catheter pressure (in the chamber) P_n corresponding to the cineangiographic time instant t_n. Utilising this data, the sequential steps adopted in the process of determining the stress dependent constitutive properties (namely the elastic modulus or stiffness) of the myocardium for each cineangiographic frame during diastole are now outlined. In other words, we will provide, herein, the methodology for the determination of the instantaneous Young's modulus and the principal stress, for each element, at each frame during diastole.

The flow chart of the analysis method is given in Fig. 6 and the major steps of the table are outlined, in detail, in the following paragraphs.

Step 1: The LV chamber outlines at two successive time intervals t_n and t_{n+1} are obtained from cineangiography. The analysis, however, is started with the frame at $t_n = t_1$ corresponding to the frame at the beginning of the diastole.

Step 2: The successive location of points on the endocardial boundary are designated by the method described below. For a particular cine derived chamber outline, the long cord or the major axis of the LV chamber is located by the line joining the apex and the mid-point of the root of the aortic valve. This long cord is divided into an equal number of segments ($m = 8$ in this analysis). Each cord perpendicular to the long cord, at each of these subdivi-

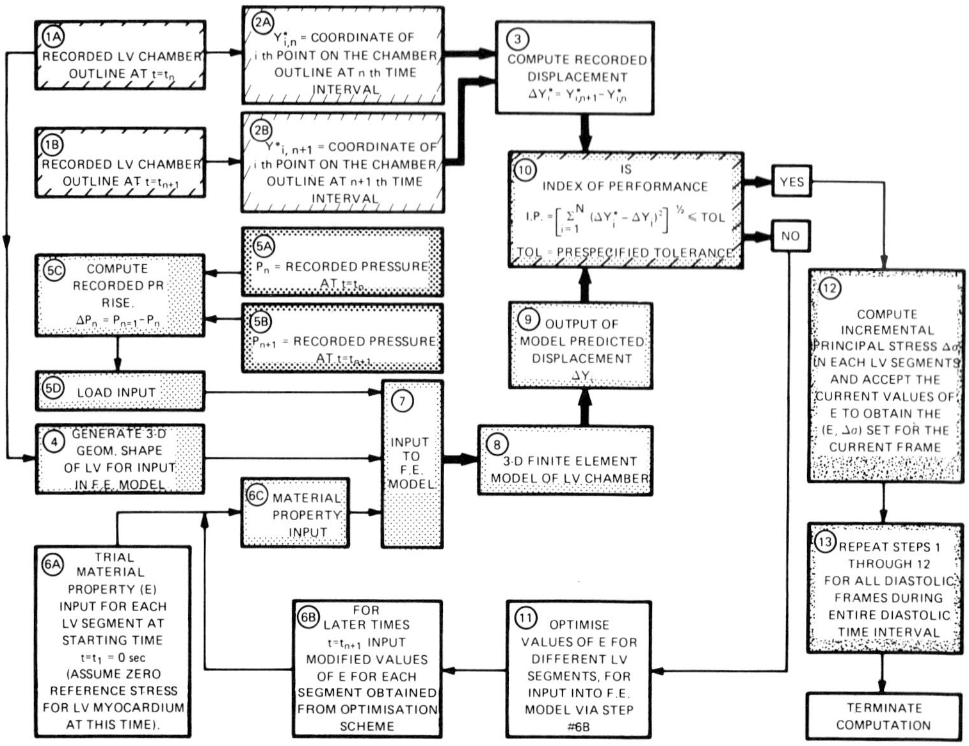

Fig. 6 The flow chart delineating the iterative procedure of matching the left ventricular model deformations with those of the monitored deformations at instant t_n, in order to determine the values of the myocardial elements' Young's modulus (E_n) and the principal stress (σ_n).

sions, intersects the endocardial chamber outline at two points. The poisition vectors of these points i at times t_n and t_{n+1} are designated by $Y^*_{i,n}$ and $Y^*_{i,n+1}$ respectively. These position vectors are designated with respect to the Newtonean reference frame whose origin is the center of the cine camera lens.

Step 3: From the above designated sequential position vectors of the point i at times t_n and t_{n+1}, $Y^*_{i,n}$ and $Y^*_{i,n+1}$, the recorded displacement vectors, during the time interval Δt ($= t_{n+1} - t_n$), are computed as:

$$\Delta Y^*_i = Y^*_{i,n+1} - Y^*_{i,n} \tag{1}$$

where i refers to the numbering of the points on the LV chamber wall outline, and n refers to the numbering of the time interval.

In Fig. 7, the LV outlines at t_n and t_{n+1} the position vectors and the displacement vector at a typical point on a segment, $m = 2$, are shown. For the sake of clarity, the displacement vector is somewhat exaggerated. This method of determining the displacement vectors of points on the chamber's endocardial outline is based on experimentally determined displacement modes of the radio opaque markers on the LV wall by Ingels et al (1975).

Step 4: A three dimensional geometrical shape of the LV at the instant t_n is reconstructed by the method suggested by Ghista and Hamid (1977). The method briefly consists of assuming an elliptical left ventricular cross-section at each segment. The major axis of this elliptical cross-section is in the anterior-posterior (A-P) projection of the cineangiogram and the minor axis is in the lateral plane. The length of the minor axis is assumed to be 90% of the length of the major axis. A typical 3-D geometrical shape of a LV chamber, to be used as an input in the Finite Element model (step 8), is shown in Fig. 8.

Step 5 (a, b, c): The recorded pressures P_n and P_{n+1} at times t_n and t_{n+1}, respectively, are used to compute the actual pressure rise,

$$\Delta P_n = P_{n+1} - P_n \tag{2}$$

during this time interval. This pressure rise is used as load input in the finite element model in step 8.

Step 6: Due to the incompressibility of the myocardium, a Poisson Ratio of 0.49 is assumed. An additional assumption of elastic isotropy requires only one other material property, i.e., elastic stiffness E, to be considered.

Computational time can be reduced by judicious choice of the trial values of the instantaneous elastic stiffness. In this regard, representative LV myocardial elastic stiffness values based on our own studies (Ghista et al, 1975a) furnish the initial trial values as input to the finite element model (of step 7).

The trial values are repeatedly modified by an optimization scheme (of step 11) until the correct value of E for each segment during a given time interval is determined.

Step 7: In this step, the three, LV parameters, namely the instantaneous three dimensional geometrical description of the LV chamber (from step 4), the instantaneous load input (from step 5) and the trial value (or the subsequent modified values) of the instantaneous elastic stiffness E (obtained from step 6) are suitably processed (as card input or tape input), for input into the instantaneous 3-D finite element model of the LV chamber.

Step 8 & 9: The 3-D finite element model of the LV chamber utilizes an eight node hexahedral type element. Each hexahedral element is assembled from five tetrahedral element with three degrees of freedom per node (Zienkiewicz, 1971). The detail formulation of the stiffness matrix for this type of element is given in Appendix B.

For the inputs presented in Step 7, the model yields the displacement vectors ΔY_i at the different points (i) on the LV chamber's inner wall. These models' predicted displacement vector outputs are now compared with the

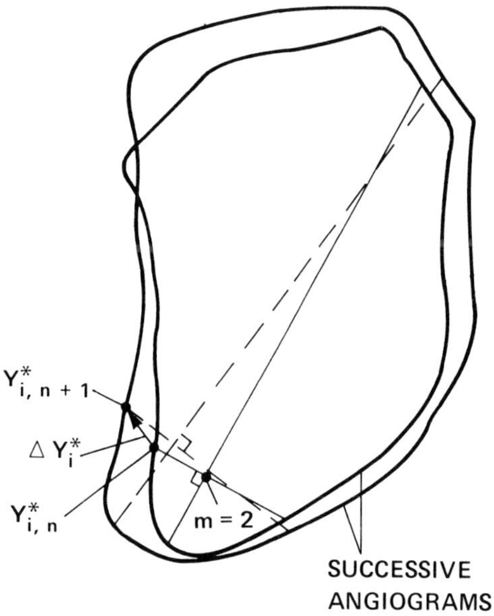

Fig. 7 Schematic illustration of the mode of translation of individual points on the left ventricular endocardial boundary (as experimentally observed by Ingles et al, 1975). This concept is employed to determine the instantaneous left ventricular endocardial displacements and velocities.

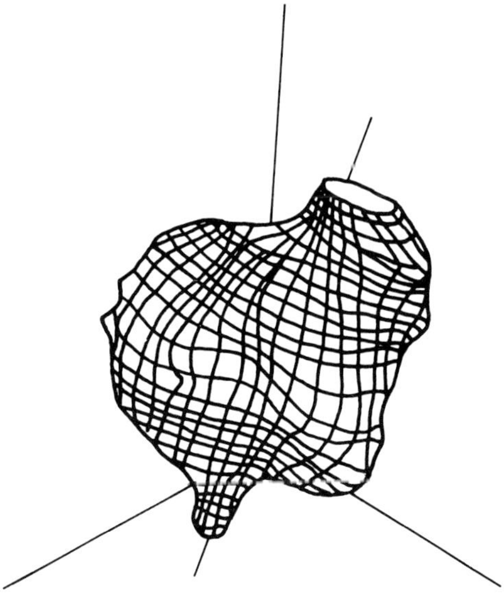

Fig. 8 A typical instantaneous 3-dimensional finite element model of the left ventricle.

recorded displacement vectors (ΔY_i^*) at the same i coordinate points obtained in Step 3. Since we are matching the incremental LV deformations for incremental chamber pressure, we are justified in employing small deformation strain and linear stress-strain characteristics in our LV finite element model (Ghista and Hamid, 1977).

Step 10: In this step, we check the difference between the model predicted incremental displacement vectors ΔY_i and the actual recorded incremental (during Δt) displacement vector ΔY_i^*. The error between these two deformation vectors is defined by

$$e = Y_1/Y_2 \tag{3}$$

where

$$Y_1 = \left[\sum_{i=1}^{N} (\Delta Y_i - \Delta Y_i^*)^2\right]^{\frac{1}{2}} / N$$

and

$$Y_2 = \left[\sum_{i=1}^{N} (\Delta Y_i^*)^2\right]^{\frac{1}{2}} / N$$

with N = total number of points monitored. If this error is not to exceed 5% (i.e., if $e = 0.05$), then the tolerance "tol" is defined by

$$\text{tol} = eNY_2 \tag{4}$$

wherein $e = 0.05$. It is hence checked whether the "Index of Performance" (I.P.) is less than "tol", i.e., if

$$\text{I.P.} = \left[\sum_{i=1}^{N} (\Delta Y_i - \Delta Y_i^*)^2\right]^{\frac{1}{2}} < \text{tol} \tag{5}$$

If the I.P. is not less than "tol", it is implied that the error "e" defined by Eq. (3) is more than 5%, and the assumed trial and/or the subsequent modified values of the elastic modulus do not represent the true values of the instantaneous myocardial stiffness. In that event, the following Step 11 is executed.

Step 11: The next set of modified values of elastic stiffness (to be used as input, in Step 6c, to the finite element model) is optimally determined by the method of "Steepest Descent". The optimization procedure developed by Nikravesh (1976) and adopted by Lin et al (1978), is given in Appendix C. Thus the steps 8 through 11 are executed repeatedly until the I.P. level, given by Eq (5), is achieved. We can now proceed to the following step.

Step 12: The strains in the elements are computed from the final set of computed deformation vectors of associated constitutive equation. From these strains, the instantaneous incremental, and hence the total stresses, are computed from the values of elastic modulus set (obtained in the above step and stored in 6b). Thus, for each segment, the maximum instantaneous principal stress and the corresponding values of the instantaneous elastic stiffness are determined.

Step 13: The above procedure, entailing steps 1 through 12, are repeated for each diastolic cineangiographic frame. The computation is terminated when all the frames for the diastolic time interval are analyzed.

B. Fluid Mechanical Analysis for the Blood Pressure Distribution in the LV Chamber

Fluid mechanical analysis of blood in the LV chamber is performed, during a small interval of time (between two successive frames) at early systole. The basic equations and the finite element formulation of this problem are presented in this section.

The purpose of this analysis is to compute the instantaneous pressure (and therefrom, if need be, the velocity distribution in the LV chamber), due to the chamber wall's instantaneous velocity field (caused by contraction) derived from two successive angiographic traces of the endocardial boundary. A two-dimensional finite element model of the fluid flow in the LV chamber is developed. The fluid is taken to be incompressible and viscous. Steady flow is assumed during the time interval between successive instants; however, the convective inertia terms are included in the analysis.

For an incompressible fluid, the density is constant, thus the continuity equation becomes,

$$u_{i,i} = 0 \tag{6}$$

The equation of motion (Navier-Stokes equation) is given by

$$\dot{u}_i + u_{i,j}u_j + 1/\rho\, P_{,i} - \nu(u_{i,jj}) = F_i \tag{7}$$

where $\dot{u}$ is the time derivative of the velocity and F_i is the body force. Since a steady state solution is sought, the $\dot{u}_i$ term is neglected, but the convective term, $u_{i,j}u_j$ contributing to the inertia, remains.

The finite element formulation and solution of Eqs. (6) and (7) yields the unknowns u_i (two velocity components) and P at the nodal points. The finite element formulation automatically includes the velocity boundary conditions on the LV chamber wall. The velocity boundary condition is obtained by taking the time derivative (employing a finite difference scheme) of the displacement field $Y^*_{i,n}$ obtained from cineangiography (see Fig. 7).

We have employed a "velocity and pressure" formulation (as opposed to stream function and vorticity formulation) because of its distinct advantage, as outlined in Taylor and Hood (1973) and Yamada (1974). The discretization of the equation of motion is also detailed in these publications.

The finite element solution employed here is an iterative solution procedure with starting trial solutions of u_i and P_i. The trial solution starts with low Reynolds Number flow and the iteration process calculates the flow at successively larger Reynolds Number using the solution at lower Reynolds Number as the initial condition for the next step. The stability of such an iterative procedure depends on the choice of a sufficiently small time increment Δt^*. A guideline suggested by Cheng (1972) is $\Delta t^* < 0.1\ (\Delta L)^2\ R_e$, where ΔL is the characteristic length(m) of the finite element and R_e is the Reynolds Number. We have found a Δt^* of about one-tenth the time interval between two successive cineangiographic frames to be quite adquate for stability. The accuracy of this methodology is also dependent on the velocity boundary conditions as determined from successive cineangiographic frames.

If the instantaneous value of the catheter pressure and the location of the catheter tip are known, then the absolute instantaneous chamber pressure distribution can be determined. However, it is not necessary from a diagnostic viewpoint to know the absolute pressure distribution, but merely the pressure distribution normalized with respect to the pressure at some point in the chamber, say at the aortic valve root in LV chamber outline. Such a normalized pressure gradient determination would not require knowledge of the chamber pressure, and hence the method would be totally noninvasive if the sequential chamber outlined here were obtained from 2-dimensional echocardiography.

IV. Results and Discussion

A. In Vivo Constitutive Property of the Normal and Pathological Myocardium During Diastole

We will first present the results of the "elastic modulus versus principal stress" constitutive property of the passive myocardium for seven subjects who are diagnosed as normal and five patients with myocardial infarction.

Fig. 9 gives the myocardial elastic modulus

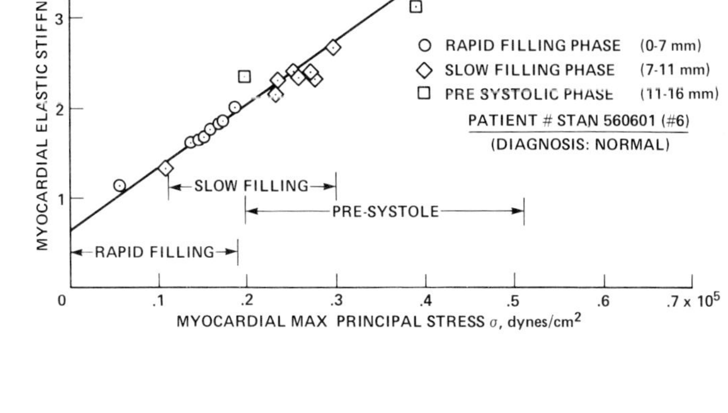

Fig. 9 Modulus (E) vs principal stress relationship for the typically normal LV, of the patient #STAN 560601.

(or stiffness) E as a function of the maximum principal stress σ, for a typical normal case. The results are presented for all the eight segments of the LV chamber, during three different phases of diastole. It may be noted that, at any given instant during diastole, the stress state as well as the elastic modulus vary over the LV myocardium. Further, as the filling progresses, the stresses and the corresponding elastic modulus increase. However, a straight line can be fitted through all the $E - \sigma$ data points, providing a measure of the stress dependent elastic modulus of the myocardium of the individual.

Fig. 10 gives the mean and the upper and lower bounds of the modulus versus principal stress relationship (i.e., of the coefficients of the E-σ relationship) of seven subjects who are clinically diagnosed as normal. These relationships (or the ranges of their coefficients) can now be utilized as bench marks to designate myocardial elements as normal or pathological.

In Fig. 11, the LV myocardial stiffness of a patient, who is clinically diagnosed to have myocardial infarctions, is plotted on this E, σ coordinate plane, superimposed on the normal bounds or domain of E versus σ (as delineated in Fig. 10. It is observed that (of the eight seg-

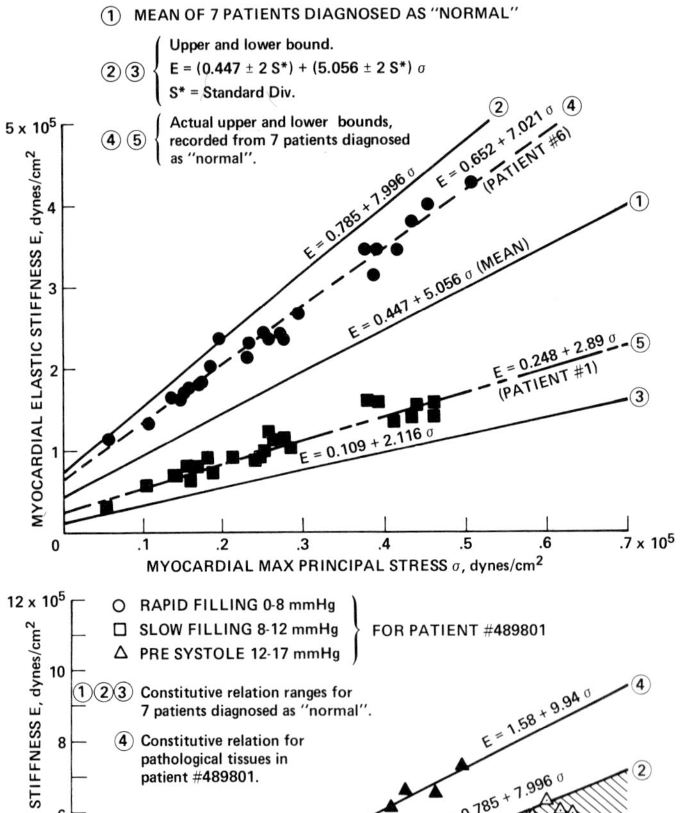

Fig. 10 The best-fit constitutive relationship for seven normal subjects, is indicated as $E = 447 \times 10^5 + 6.056\sigma$. In order to designate the normal myocardium's (E,σ) domain, the lower and upper bounds for the normal myocardium (whose mean property is $E = 447 \times 10^5 + 6.0565\sigma$) are designated at two standard deviations from this mean constitutive relationship. Of the seven normal subjects, the LV elements of patients #1 and #6 just lie within these lower and upper bounds of the normal domain, while the LV elements of the remaining five subjects lie well within this normal domain.

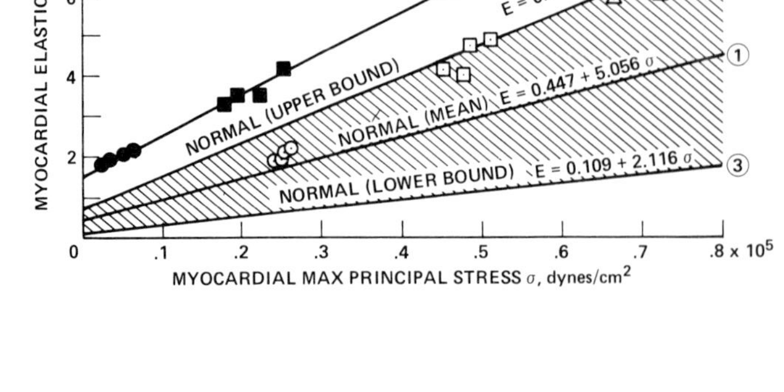

Fig. 11 Note that some pathological elements of a subject (#489801) having myocardial infarction, fall outside the normal domain designated by the lower and upper bounds of Fig. 9; the remaining (four) elements fall within the normal domain.

ments of the LV chamber) the elastic E-σ properties of four of his LV segments fall in the normal range, while the other four segments have an abnormally elevated (stress dependent) myocardial stiffness. The segments which are thus diagnosed to be pathological are numbered, so that they may then be properly depicted on the angiographic trace. Through those (E,σ) coordinates that fall beyond the normal E-σ domain, a linear E-σ relationship is fitted to characterize the constitutive property of the pathological myocardium. Fig. 12 presents the computer predicted normal-pathological myocardial elements of a typical patient.

Fig. 13 gives the mean, upper and lower bounds of all the values of (E,σ) of the five patients that fall outside the normal domain defined by Fig. 10. Since the domains of the normal and pathological elements are distinctly separate, one can clearly distinguish normal myocardial elements from those with coronary artery disease on the basis of their constitutive property coordinate sets (E,σ) falling within or outside the normal domain defined on the (E,σ) coordinate plane in Fig. 10.

The accuracy and diagnostic sensitivity of our methodology for determination of diastolic myocardial constitutive property can be improved by analyzing successive angiographic frames at very small time intervals. The analysis of cineangiographic frames at smaller time intervals will enable closer approximation of

PATIENT #474701-1

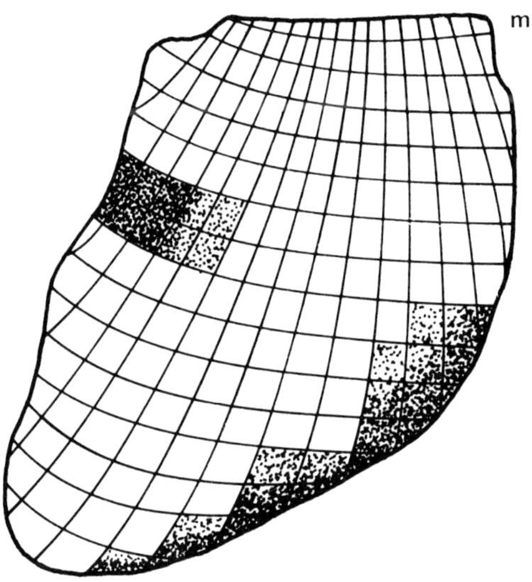

Fig. 12 Schematic depiction of the pathological elements on the planar finite element discretization of the left ventricular chamber.

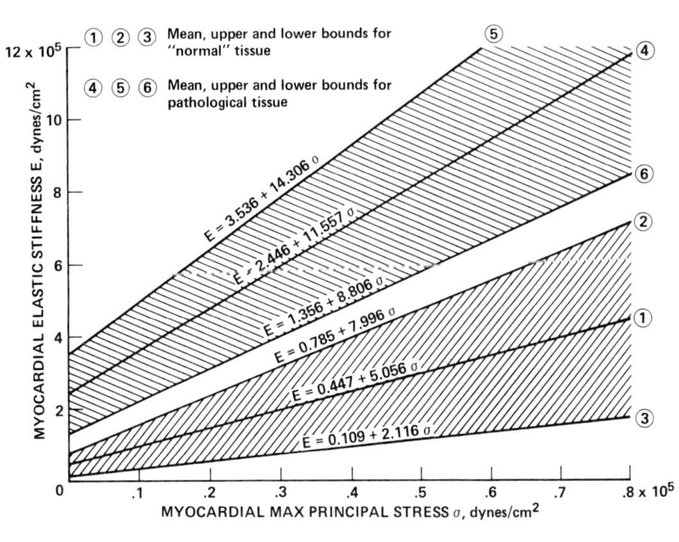

Fig. 13 For all those myocardial elements that do not fall within the designated normal domain, the best-fit, lower bound and upper bound constitutive relationships are determined as $E = 2.446 \times 10^5 + 11.557\sigma$, $E = 1.356 \times 10^5 + 9.8065\sigma$ & $E = 3.536 \times 10^5 + 14.306\sigma$. Thus, the pathological myocardial domain is determined and designated to be distinctly separate from the "normal" domain. Thus, by determining the instantaneous value of (E,σ) of a LV myocardial element, and plotting the coordinate pair on the above $E - \sigma$ constitutive plane, we can determine if that element is normal or diseased.

the actual nonlinear relationship of E versus σ, (by its linearization through determining the constitutive set E, σ) at discrete times.

Towards a Noninvasive Procedure for Detection of Normal and Pathological Myocardial Elements

The formulation presented earlier (in Section III-A) enables us to obtain the distribution of the regional elastic stiffness E with respect to associated maximum principal stress σ, provided (i) the LV geometry wall thickness (t) and the chamber pressure (p) are known. However, in clinical situations, data on LV chamber pressure can only be obtained by catheterization, whereas the LV wall thickness is difficult to record. Now, whereas the actual values of the myocardial stress σ and the stiffness E depend on the actual values of p and t, the slope of the E vs. σ curve (i.e., $dE/d\sigma$) will be independent of p and t. In other words, the slope of the constitutive relation defined by the straight line fit of the E vs σ data is independent of the values of p and t, even though the actual points on this E vs σ straight line would depend on the actual values of p and t.

In order to understand the basis of the above concept, we note that in the analysis for determination of the optimum values of the elements' stiffnesses (E), we are forcing the displacement field solution for the finite element analysis $\{\delta\}$ to a predetermined strain field $\{\varepsilon^*\}$ defined by the predetermined deformation field $\{\delta^*\}$ obtained from the cineangiogram, in the [C] matrix (Appendix B). Thus $\{\varepsilon\}$ obtained from this analysis will be independent of p and t, and will depend only on the deformation field $\{\delta^*\}$ given by the cineangiogram. Thus, the incremental strain field $\{\Delta\varepsilon\}$ from frame to frame will also be independent of Δp and t. Now, since the incremental strain field $\{\Delta\varepsilon\}$ corresponding to $\{\Delta\sigma\}$ is given by $d\sigma/dE$ (the inverse of slope of E-σ curve), it follows that $d\sigma/dE$ is independent of p and t.

It can also be noted, from the Appendix Eq. B-9, that $d\sigma/dE$ is dependent on ν and the known value $\varepsilon = \varepsilon^*$. Thus $d\sigma/dE$ is independent of p and t. In other words, the slope of the E-σ curve, $E = a + b\sigma$, for the myocardial tissue (in Figs. 9, 10, 11, 13) can be determined without knowledge of the precise values of the chamber wall thickness and pressure. However, the intercept "a" of this E-σ curve ($E = a + b\sigma$) has the same unit as stiffness, and thus cannot be accurately evaluated without knowledge of the actual values of p and t.

Since the parameter "a" (the intercept) may be physically interpreted as the elastic stiffness at zero stress, it is actually the elastic stiffness of the myocardium during the time interval of the very first frame from the beginning of diastole. Assuming linearity of pressure rise during this diastole, the physiological range for p during the time interval of the first frame at early diastole will be extremely small, particularly if high speed cineangiography is used.

As an example, if we assume the range of diastolic pressure rise to be almost 10 to 15 mmHg and if there are, say, 15 cineangiographic frames during diastole, then the range of pressure rise during the first frame could be between ($10/15 = 0.67$) to ($15/15 = 1.0$) mmHg, yielding a 33% variation in the pressure value. If the actual chamber thickness t is determined noninvasively (e.g. by echocardiogram), then one could expect a maximum variability of 33% in the estimation of the intercept "a", due to above mentioned 33% variability in the value of the assumed pressure. However, an observation of our results (Fig. 13), indicates that based on the values of "a" determined from the values of p and t, there is at least that much variability within each normal or infarcted group, but of course an order of magnitude difference between the "a" values of normal and infarcted groups (also, refer the table below).

The following table provides the variabilities in the computed values of "a", for normal and pathological myocardiums.

	Mean value of "a" in dyne/cm^2	% Variation
Normal group	0.447×10^5	±75.6%
Infarcted group	2.46×10^5	±44.6%
Between mean of normal and infarcted		±447%

Note that the ranges of "a" values of normal and pathological are $0.3 \times 10^5 - 0.6 \times 10^5$ dyne/cm^2 and $2 \times 10^5 - 3 \times 10^5$ dyne/cm^2. Thus, 33% variability in the value of "a" due to erroneous estimation of the pressure rise during the time interval of the first diastolic frame would still yield maximal normal and infarcted ranges of $0.25 \times 10^5 - 0.7 \times 10^5$ dyne/cm^2 and $1.7 \times 10^5 - 3.5 \times 10^5$ dyne/cm^2. Thus, even by assuming an arbitrary but reasonable value of the pressure rise during the time interval of the first frame, the ranges of normal and infarcted tissues values of the parameter "a" are distinct.

Thus a reasonable assumption of the initial diastolic pressure would enable us to obtain an "a" value that would be at most 33% off from the actual "a" value and still enable us to clearly categorize the myocardium as normal or infarcted. Of course, assumptions of the arbitrary value of the end diastolic pressure and its linear variation would still yield the precise value of the slope parameter b of the constitutive relation $E = a + b\sigma$. Since this procedure of determining the values of the constitutive parameters "a" & "b" from an assumed value of end-diastolic pressure can distinguish between normal and pathological myocardium, we have laid down the foundations of a totally noninvasive procedure for detecting pathological myocardial elements.

B. Pressure Distribution of Blood Flow in the Left Ventricular Chamber During Systole

The results of the fluid mechanical analysis are in qualitative agreement with the concept of the intraventricular pressure gradient promoting LV chamber ejection and the experimental findings, in Ray et al (1977), as demonstrated in Figs. 2 through 5. Figures 14 (a & b) and 15 (a & b) show the distribution of the intraventricular pressure fields for two patients with myocardial infarction, before and after coronary bypass surgery, at instants corresponding to the cineangiographic traces at the start of ejection. The pre-surgery results show a large area in the LV chamber having a "reversed" pressure gradient (i.e., not directed towards the aortic valve) and thus not conducive to efficient ejection. After surgery, the zone of "reversed" pressure gradient still exists, but the domain of "favorable" pressure gradient is much larger. Corresponding improvement in ejection fraction (obtained from computation of ejected volume and end diastolic volume from cineangiography) can also be noted.

V. Concluding Comments

1. This chapter provides, for the first time, the values of the in vivo elastic constitutive (E vs σ) properties of normal and pathological myocardial elements. In so doing, it also paves the way for detection of the diseased myocardial elements; a stepwise procedure is provided for this purpose.

2. The computed blood pressure distributions in the LV chamber (of Figs. 14 and 15) graphically illustrate the efficiency of LV myocardial contraction to produce effective (high ejection fraction facilitating) pressure gradient. The set of observations (depicted in Figs. 14 and 15) based on the chamber pressure distribution provides an integrated index of LV myocardial contractility, and (as shown herein) enables us to graphically study the efficacy of coronary bypass surgery.

VI. Appendix A: Experimental Findings of the Intraventricular Blood Pressure Gradient

The pressure recordings were obtained by two Miller ® catheter-tipped micropressure transducers PC 370 and PC 350 introduced into the left ventricle through the femoral and left carotid arteries. These transducers were radioopaque under the flouroscope. One of these was positioned near the apex and the other proximal to the aortic valve. The lumen catheter was used to inject the radioopaque dye so that the LV chamber could be seen simultaneously with the catheter tipped transducer located in situ.

Let us elucidate the concept of a pressure gradient in the LV chamber, during isovolumic contraction and ejection with the following examples. If a fluid-filled container is accelerated, it is well known that a dynamic transient pressure gradient exists between the leading and trailing edges of the container. Similarly, if a fluid-filled container is impacted, a dynamic pressure gradient will result because of the container acceleration and the propagation of the impact deformation. Since different parts of the myocardium contract at different times, it is natural for a gradient to exist during isovolumic contraction or ejection. Moreover, a positive pressure gradient developed during isovolumic contraction towards the aortic valve region will result in a more efficient ejection than that of a uniform pressure rise in the chamber.

A point that could be raised is that the high pressures observed in the transducer close to the apex is due to the so-called "transducer entrapment", due to contact with the apex myocardium during the ejection phase. It is our observation that if the transducer is being impacted, squeezed, or "trapped" by the myo-

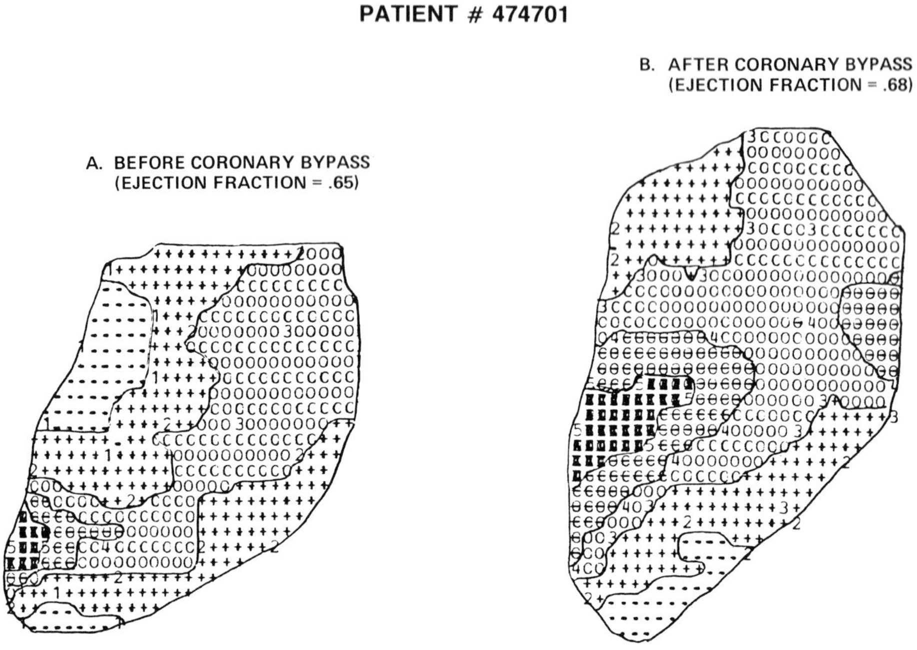

NONDIMENSIONAL PRESSURE		SYMBOL	% LV CHAMBER AREA AT PRESSURE LEVEL		COMMENTS (A NEO INTRINSIC MEASURE OF IMPROVED CONTRACTIVITY AFTER CORONARY BYPASS)
LEVEL	VALUE (P/P_0) (P_0 = 1, AT AV BOUNDARY OF LV CHAMBER)		BEFORE BYPASS	AFTER BYPASS	
1	0.80 – 0.90	- - - - - - - - -	12.7 } = 56.3	5.8 } = 32.4	DECREASE OF ZONE OF LOW PRESSURE LEVEL (1 & 2) BY 23.9% AFTER BYPASS
2	0.90 – 0.98	+ + + + + + + + +	43.6	26.6	
3	0.98 – 1.02	0 0 0 0 0 0 0 0 0	38.0 = 38.0	45.1 = 45.1	INCREASE OF ZONE OF PRESSURE LEVEL (3) BY 7.1% AFTER BYPASS
4	1.02 – 1.08	θ θ θ θ θ θ θ θ θ	3.8 } = 5.7	17.4 } = 22.5	INCREASE OF ZONE OF HIGH PRESSURE LEVELS (4 & 5) BY 16.8% AFTER BYPASS
5	1.08 – 1.16	■ ■ ■ ■ ■ ■ ■ ■ ■	1.9	5.1	

Fig. 14 The relative pressure distribution in the left ventricular chamber (just prior to ejection) of a subject #474701 (having a myocardial infarction) before and after coronary bypass surgery.

PATIENT # 507001

PATIENT #507001-1 PATIENT #507001-2

B. AFTER CORONARY BYPASS
(EJECTION FRACTION = .75)

A. BEFORE CORONARY BYPASS
(EJECTION FRACTION = .68)

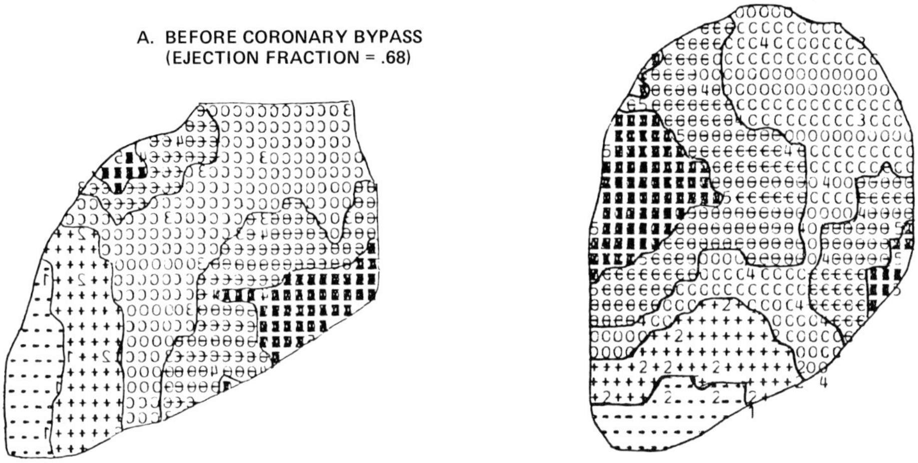

NONDIMENSIONAL PRESSURE		SYMBOL	% LV CHAMBER AREA AT PRESSURE LEVEL		COMMENTS (A NEO INTRINSIC MEASURE OF IMPROVED CONTRACTIVITY AFTER CORONARY BYPASS)
LEVEL	VALUE (P/P_0) (P_0 = 1, AT AV BOUNDARY OF LV CHAMBER		BEFORE BYPASS	AFTER BYPASS	
1	0.84 – 0.92	- - -	10.28 } = 36.41	7.55 } = 20.73	DECREASE OF ZONE OF LOW PRESSURE LEVELS (1 & 2) BY 15.68% AFTER BYPASS
2	0.92 – 0.99	+ + +	26.13	13.18	
3	0.99 – 1.02	0 0 0	38.89 = 38.89	37.47 = 37.47	
4	1.02 – 1.10	θ θ θ	15.75 } = 24.7	26.93 } = 41.8	INCREASE OF ZONE OF HIGH PRESSURE LEVELS (4 & 5) BY 17.1% AFTER BYPASS
5	1.10 – 1.17	■ ■ ■	8.95	14.87	

Fig. 15 The relative pressure distribution in the left ventricular chamber (just prior to ejection) of a subject #507001 before and after coronary bypass surgery.

cardium, the pressure pulses will have marked deviations, in the form of either (a) a marked spike due to the catheter-tip (where the strain gauge transducer is located) being "squeezed" between the papillary muscle and the myocardium, or (b) a heavily damped pressure signal if the catheter tip is embedded in the trabeculae carneae.

The results presented, although somewhat preliminary in nature, suggest that (a) an intraventricular pressure gradient exists during systole, (b) that this pressure gradient is altered with induced changes in the LV myocardial perfusion, and (c) that correspondingly this pressure distribution is characteristic of LV performance.

Very little attention has hitherto been paid to the chamber hemodynamic aspects of the left ventricular function. An understanding of the nature of blood flow and pressure distribution and its correlation with the myocardial pathology could provide more illustrative and insight provoking measures of (normal and pathological) left ventricular functional efficiency. Evidence of this is provided in section IV of this chapter.

VII. Appendix B: 3-Dimensional Finite Element Model of the LV for LV-Mechanomyocardiography

The 3-dimensional left ventricular chamber is reconstructed by a number of eight node hexahedral type elements, each of which is composed of five node tetrahedral elements (Zienkiewicz 1971), with nodes i, j, k, l. The coordinates of a typical nodal point i of a tetrahedral element are given by (x_i, y_i, z_i), whereas a typical point p inside the tetrahedron is represented by the coordinate x, y, z).

The relationship between the global cortesian coordinates and the local "Natural coordinates" is given by:

$$\begin{bmatrix} x \\ y \\ z \\ 1 \end{bmatrix} = \begin{bmatrix} x_i & x_j & x_k & x_l \\ y_i & y_j & y_k & y_l \\ z_i & z_j & z_k & z_l \\ 1 & 1 & 1 & 1 \end{bmatrix} \begin{bmatrix} L_i \\ L_j \\ L_k \\ L_l \end{bmatrix} \qquad \text{(B-1)}$$

wherein

$$\begin{bmatrix} L_i \\ L_j \\ L_k \\ L_l \end{bmatrix} = \frac{1}{6v} \begin{bmatrix} a_i & b_i & c_i & d_i \\ a_j & b_j & c_j & d_j \\ a_k & b_k & c_k & d_k \\ a_l & b_l & c_l & d_l \end{bmatrix} \begin{bmatrix} 1 \\ x \\ y \\ z \end{bmatrix} \qquad \text{(B-2)}$$

v is the Volume of the tetrahedron, and (a_i, b_i, c_i and d_i) are expressed in terms of their nodal coordinate as

$$a_i = \begin{bmatrix} x_j & y_j & z_j \\ x_k & y_k & z_k \\ x_l & y_l & z_l \end{bmatrix}; \quad b_i = \begin{bmatrix} 1 & y_j & z_j \\ 1 & y_k & z_k \\ 1 & y_l & z_l \end{bmatrix} \qquad \text{(B-3)}$$

$$c_i = -\begin{bmatrix} x_j & 1 & z_j \\ x_k & 1 & z_k \\ x_l & 1 & z_l \end{bmatrix}; \quad d_i = -\begin{bmatrix} x_j & y_j & 1 \\ x_k & y_k & 1 \\ x_l & y_l & 1 \end{bmatrix}$$

In order to maintain the appropriate signs of the above equations, the right hand system of numbering the nodes is maintained throughout the program.

Displacement Field of the Tetrahedral Element:

The assumed displacement field of the tetrahedral element is obtained by linear interpolation between nodes, and is given in terms of the natural coordinates, given in Eq. (B-2), by:

$$\{\delta\} = \begin{Bmatrix} u \\ v \\ w \end{Bmatrix} = \begin{Bmatrix} \sum_{n=i}^{l} N_n(x,y,z)u_n \\ \sum_{n=i}^{l} N_n(x,y,z)v_n \\ \sum_{n=i}^{l} N_n(x,y,z)w_n \end{Bmatrix} = \begin{Bmatrix} [N] & \{u\} \\ [N] & \{v\} \\ [N] & \{w\} \end{Bmatrix} \qquad \text{(B-4)}$$

$$\text{where } [N] = [L_i L_j L_k L_l],\ \{u\} = \begin{Bmatrix} u_i \\ u_j \\ u_k \\ u_l \end{Bmatrix}, \text{ etc.} \qquad \text{(B-5)}$$

Strain-Displacement Relation

The usual small strain-displacement relation is employed, namely

$$\{\varepsilon\} = [B]\{\delta\} \qquad \text{(B-6)}$$

where

$$[B] = [B_i B_j B_k B_l] \qquad \text{(B-7)}$$

and the typical term B_i is given by

$$[B_i] = \begin{bmatrix} L_{i,x} & 0 & 0 \\ 0 & L_{i,y} & 0 \\ 0 & 0 & L_{i,z} \\ L_{i,y} & L_{i,x} & 0 \\ 0 & L_{i,z} & L_{i,y} \\ L_{i,z} & 0 & L_{i,x} \end{bmatrix} = \frac{1}{6v} \begin{bmatrix} b_i & 0 & 0 \\ 0 & c_i & 0 \\ 0 & 0 & d_i \\ c_i & d_i & 0 \\ p & d_i & c_i \\ d_i & 0 & b_i \end{bmatrix} \qquad \text{(B-8)}$$

Stress-Strain relation

Assuming linear elasticity and isotropy, the stress-strain relation for an element is given by:

$$\{\sigma\} = [C]\{\varepsilon\} \qquad \text{(B-9)}$$

where the material matrix $[C]$ is given by:

$$[C] = \begin{bmatrix} C_{11} & C_{12} & C_{13} & 0 & 0 & 0 \\ & C_{22} & C_{23} & 0 & 0 & 0 \\ & & C_{33} & 0 & 0 & 0 \\ & \text{Sym} & & C_{44} & 0 & 0 \\ & & & & C_{55} & 0 \\ & & & & & C_{66} \end{bmatrix} \qquad \text{(B-10)}$$

and,

$$C_{11} = C_{22} = C_{33} = \frac{E(1 - v)}{(1 + v)(1 - 2v)}$$

$$C_{12} = C_{13} = C_{23} = \frac{E}{(1 + v)(1 - 2v)}$$

$$C_{44} = C_{55} = C_{66} = \frac{E}{2(1 + v)} \qquad \text{(B-11)}$$

where the Poisson's ratio (v) is taken equal to 0.49.

Minimum Potential Energy Principle:

The potential energy functional is given by:

$$\pi(u,v,w) = U - W \qquad \text{(B-12)}$$

where U and W are respectively the strain energy and work done due to the pressure on the inner surface of the LV chamber. They are given by $U = \frac{1}{2}\int_{vol} \{\sigma\}^T \{\varepsilon\}\, dv$ and $W = \int_s \{\bar{p}\}^T \{\bar{\delta}\}\, ds$. By substituting (B-6) and (B-9) therein, we obtain

$$\pi = \frac{1}{2}\int_{vol} \{\delta\}^T [B]^T [C] [B] \{\bar{\delta}\}\, dv - \int_s \{\bar{p}\}^T \{\bar{\delta}\}\, ds \qquad \text{(B-13)}$$

where $\{\bar{p}\}$ provides the components of the pressure normal to the endocardial surface. For the minimum value of the above energy functional, we have

$$\delta\pi = \sum_{e=1}^{n} \delta\pi^{(e)}(u, v, w) = 0 \qquad \text{(B-14)}$$

and since

$$\delta\pi^{(e)}(u,v,w) = \sum_{i=1}^{m} \frac{\partial \pi^{(e)}}{\partial u_i}\delta u_i + \sum_{i=1}^{m} \frac{\partial \pi^{(e)}}{\partial v_i}\delta v_i + \sum_{i=1}^{m} \frac{\partial \pi^{(e)}}{\partial w_i}\delta w_i$$

we have

$$\frac{\partial \pi^{(e)}}{\partial u_i} = \frac{\partial \pi^{(e)}}{\partial v_i} = \frac{\partial \pi^{(e)}}{\partial w_i} = 0 \qquad \text{(B-15)}$$

In the above Eqs. (B-14) and (B-15), the symbol "e" refers to the element. The symbols m and n represent, respectively, the total number of nodes in a given element and the total number of elements in the system.

On differentiating Eq. (B-13), we obtain the discretised equation for an element (at a given node i) as

$$\left\{\frac{\partial \pi^{(e)}}{\partial \delta}\right\} = \left\{\begin{matrix} \dfrac{\partial \pi^{(e)}}{\partial u_i} \\ \dfrac{\partial \pi^{(e)}}{\partial v_i} \\ \dfrac{\partial \pi^{(e)}}{\partial w_i} \end{matrix}\right\} = \int_{vol} [B_i^e]^T [C^e] [B_i^e] \{\overline{\delta^e}\}\, dv - \int_s [N_i^e] \{\overline{p^e}\}\, ds = 0 \qquad \text{(B-16)}$$

leading to

$$\int_{vol} [B_i^e]^T [C^e] [B_i^e] \{\bar{\delta}_i^e\} \, dv$$

$$= \int_s [N_i^e] \{\bar{p}^e\} \, ds \qquad \text{(B-17)}$$

Note that for the nodes which do not lie on the endocardial wall of the chamber, the right hand side of Eq. (B-17) is automatically zero.

Eq. (B-17) may be rewritten as

$$[K_i^e] \{\bar{\delta}_i^e\} = \{F_i^e\} \qquad \text{(B-18)}$$

where

$$[K_i^e] = \int_{vol} [B_i^e]^T [C^e] [B_i^e] \, dv \qquad \text{(B-19)}$$

and

$$\{F_i^e\} = \int_s [N_i^e] \{\overline{p_i^e}\} \, ds \qquad \text{(B-20)}$$

Once all the element stiffness matrices are formed according to Eqs. (B-19), they are assembled to form the overall stiffness matrix, (whose size would be $[3m \times 3m]$ where m is the total number of nodal points) given by

$$K_{ij} \bar{\delta}_j = F_j \qquad \text{(B-21)}$$

where $i = j = 3m$, m being the total number of nodal points.

The solution of Eq. (B-21) yields the nodal deformations $\bar{\delta}_j$ predicting the LV chamber deformation, wherefrom the nodal displacements at the endocardial nodes (denoted by $\Delta Y_{i,n}^m$ in the text) are obtained. These in turn yield the chamber geometry at $(n - 1)$ instant, and hence the change in chamber volume ΔV_n. The calculated ΔV_n can be employed to alternately determine the myocardial elements' E values by matching the model and monitored values of the instantaneous changes in chamber volumes (instead of the endocardial deformations).

Optimization Procedure

It may be noted, from Eq. (B-8), that once the element configurations are specified, the (B) matrix is completely specified. Also, in Eqs. (B-10) and (B-11), the only variable in the material matrix (C) is the elastic stiffness E. Hence, Eq. (B-19) may be rewritten as:

$$[K_i^e] = E^e \int_{vol} [B_i^e]^T [C'^e_i] [B_i^e] \, dv \qquad \text{(B-22)}$$

where

$$[C'^e_i] = \frac{1}{E^e} [C_i^e]$$

Likewise, the reconstructed endocardial surface at the previous instant $n - 1$ (from the displacements $\Delta Y_{i,n}^m$ on the endocardial surface), and hence the change in chamber volume V_n (between instants n and $n - 1$), are functions of the moduli of the elements.

VIII. Appendix C: The Optimization Scheme for Determining the Elastic Stiffness

An optimization program was used in conjunction with the LV finite element analysis program to find the best overall fit between the model and actual recorded deformations.

The program used here has fast convergence and better accuracy in searching for the minimum when compared to classical optimization schemes. It consists of two phases:

Phase I:

This phase assumes the function to be minimized is quadratic and has, in matrix notation, the form

$$f(x) = f(0) + a^T x + (\tfrac{1}{2}) x^T G x, \qquad \text{(C-1)}$$

where

- x = vector of variables $(n \times 1)$,
- $f(0)$ = the value of the function $f(x)$ at point $x = 0$,
- a = vector of constants $(n \times 1)$,
- G = $(n \times n)$ symmetric positive definite matrix of 2nd order partial derivatives,
- n = number of variables.

The gradient vector (the first order partial derivative vector) of $f(x)$ is in the form

$$g(x) = a + Gx \qquad \text{(C-2)}$$

At the minimum of $f(x)$ we have

$$0 = a = Gh, \qquad \text{(C-3)}$$

where h is the value of x at the minimum.

Using the above algorithm, the function f(x) is first calculated at the point x_i, which is

$$f(x_i) = f(0) + a^T x_i + (\tfrac{1}{2}) x_i^T G x_i \quad \text{(C-4)}$$

At the point $x_{i+1} = x_i + \Delta x$, (C-1) becomes

$$f(x_{i+1}) = f(0) + a^T x_{i+1} + (\tfrac{1}{2}) x_{i+1}^T G x_{i+1} \quad \text{(C-5)}$$

Substituting (C-4) into (C-5) and making use of (C-2) yields, after some manipulations,

$$f(x_{i+1}) = f(x_i) + g_i^T \Delta x + (\tfrac{1}{2}) \Delta x^T G \Delta x \quad \text{(C-6)}$$

From Eq. (C-6) and the vector g, the matrix G can be calculated by a centered difference method. This calculation requires $n(n + 3)/2 + 1$ function evaluations or $n(n + 7)/2 + 1$ function evaluations for a better approximation. After evaluations of g and G, the following equation yields, h, the minimum of f(x).

$$h = x_i - G^{-1} g_i$$

In most practical cases, the function f(x) is not in the quadratic form of (1), therefore this method will yield the minimum only after a finite number of iterations.

Phase II:

In the second phase of this program, the given function is evaluated along the best path from the generic point x to h, the minimum, based on the information gained in Phase I. The best path means the optimal direction for parametric variation to reach the minimum. Along this directional line, a second order function is first used for interpolation to find the desired minimum. If it is not adequate, then either a third- or fourth-order function will be adopted to fit the function f(x). The local minimum thus found will then be the starting point for the next iteration and so on until the index of performance is within the pre-set error tolerance.

To mate with the finite element analysis program, the optimization program outlined above was modified. A schematic flow diagram of the combined program is shown in Fig. 6. The technique adopted required the finding of the minimum of the index of performance. The goal was achieved by varying the material constants according to the method of Newton-Raphson to alter the deformation responses of the finite element model.

IX. Appendix D: 2-Dimensional Finite Element Formulation for LV-Hemodynamics (LV-Mechanocardiopressography)

The procedure entails dividing the fluid region in the left ventricular chamber into triangular finite elements, to formulate the governing equations, (a) applying the appropriate velocity and pressure boundary conditions, and (b) computer solution of the resulting set of equations, to yield the pressure and velocity distribution within the LV chamber during the entire cardiac cycle.

Finite Element Formulation

The two dimensional flow domain (Ω) of the LV chamber bounded by the endocardial cineangiographic trace (C) is made up of a collection of 6-noded triangular elements. A triangular finite element (with 6 nodes) for a viscous fluid, is adopted herein. The nodal variables for the element are (1.) the two velocity components u_n, v_n at the six nodes ($n = 1, \ldots, 6$), and (2.) the pressure p at the three nodes (with $p = 4, 5, 6$). The velocity and pressure fields interpolated over the domain of a single triangular element are given by

$$\begin{aligned} u^{(e)} &= \sum_{n=1}^{6} N_n u_n = [N]\{u\} \\ v^{(e)} &= \sum_{n=1}^{6} N_n v_n = [N]\{v\} \quad \text{(D-1)} \\ \textit{and } p^{(e)} &= \sum_{p=4}^{6} N_p^* p_p = [N^*]\{p\} \end{aligned}$$

Note that while the interpolation for the velocities are quadratic, that for the pressure is linear as shown in Eq. (D-1). The relationships for N_n and N_p^*, for quadratic interpolation functions for velocity and linear interpolation function for pressure, are given below.

$$\begin{aligned} N_1 &= L_1^2 - L_1(L_2 + L_3) \\ N_2 &= L_2^2 - L_2(L_3 + L_1) \\ N_3 &= L_3^2 - L_3(L_1 + L_2) \\ N_4 &= 4\,L_1 L_2 \\ N_5 &= 4\,L_2 L_3 \\ N_6 &= 4\,L_3 L_1 \end{aligned} \quad \text{(D-2)}$$

$$\text{and } N_p^* = L_p \quad \text{(D-3)}$$

where L_i in natural coordinates is given by

$$L_i = \frac{a_i + b_i x + c_i y}{2\Delta} \quad \text{(D-4a)}$$

$$2\Delta = 2x \text{ (area of triangle)} = \begin{vmatrix} 1 & X_1 & Y_1 \\ 1 & X_2 & Y_2 \\ 1 & X_3 & Y_3 \end{vmatrix} \qquad \text{(D-4b)}$$

and $a_i = X_j Y_m - X_m Y_j$; $b_i = Y_j - Y_m$;
$C_i = X_j - Y_m$ (D-4c)

The solution procedure starts with a trial solution (u_n, v_n, P_n), which represents some approximate solution of a low Reynold's Number flow pattern. The governing continuity and momentum equations of steady flow are given by

$$u_{i,i} = 0 \qquad \text{(D-5)}$$

and

$$u_{i,j}u_j + \frac{1}{\rho}P_{,i} - \frac{\mu}{\rho}u_{i,jj} = 0 \qquad \text{(D-6)}$$

When the element trial functions are introduced therein and the Galerkin criterion is applied (Zienkiewicz 1971), the element stiffness matrix is obtained as

$$\left[\begin{array}{c:c:c} [K_1] & [0_1] & [K_2] \\ \hdashline [0_1] & [K_1] & [K_3] \\ \hdashline [K_4] & [K_5] & [0_2] \end{array}\right] \begin{Bmatrix} u_1 \\ \vdots \\ u_6 \\ \hline v_1 \\ \vdots \\ v_6 \\ \hline P_1 \\ P_2 \\ P_3 \end{Bmatrix} = \begin{Bmatrix} R_1 \\ \vdots \\ R_6 \\ \hline R'_1 \\ \vdots \\ R'_6 \\ \hline 0 \\ 0 \\ 0 \end{Bmatrix} \qquad \text{(D-7)}$$

where

$[K_1]$ is a 6 × 6 matrix with a typical

$$K_{ij} = \iint_{\Omega^{(e)}} \left[\frac{\mu}{\rho}(N_{i,x}N_{j,x} + N_{i,y}N_{j,y}) - u_n^{(e)}N_iN_{j,x} - v_n^{(e)}N_iN_{j,y}\right]\mathrm{d}x\mathrm{d}y \qquad \text{(D-8a)}$$

$[K_2]$ is a 3 × 6 matrix with a typical

$$K_{ij} = -\int_{\Omega^{(e)}}(1/\rho)N_{i,x}N_j^* dxdy \qquad \text{(D-8b)}$$

$[K_3]$ is a 3 × 6 matrix with a typical

$$K_{ij} = -\int_{\Omega^{(e)}}(1/\rho)N_{i,y}N_j^* \mathrm{d}x\mathrm{d}y \qquad \text{(D-8c)}$$

$[K_4]$ is a 6 × 3 matrix with a typical

$$K_{ij} = \int_{\Omega^{(e)}} N_i^* N_{j,x}\mathrm{d}x\mathrm{d}y \qquad \text{(D-8d)}$$

$[K_5]$ is a 6 × 3 matrix with a typical

$$K_{ij} = \int_{\Omega^{(e)}} N_i^* N_{j,y}\mathrm{d}x\mathrm{d}y \qquad \text{(D-8e)}$$

$[0_1]$ is a 6 × 6 null matrix

$[0_2]$ is a 3 × 3 null matrix

$R_i = 0$ if the node is internal and

$R_i = N_i\bar{X}^*\mathrm{d}s$ if the node is on the boundary (D-8f)

$R'_i = 0$ is the node is internal and

$R'_i = \int_c N_i\bar{Y}^*\mathrm{d}s$ if the node is on the boundary (D-8g)

Spatial derivatives of N required to compute K_{ij} could easily be obtained by substituting Eqs. (D-4a), (D-4b) and (D-4c) in Eqs. (D-2) and (D-3), and taking the proper derivatives, i.e.,

$$\left.\begin{aligned} N^*_{i,x} &= L_{i,x} = b_i/2\Delta \\ N^*_{i,y} &= L_{i,y} = c_i/2\Delta \end{aligned}\right\} \qquad \text{(D-9a)}$$

and typically

$$\begin{aligned} N_{4,x} &= \frac{\mathrm{d}}{\mathrm{d}x}(4L_1L_2) \\ &= \frac{\mathrm{d}}{\mathrm{d}x}[(4/2\Delta)(a_1 + b_1x + c_1y) \\ &\qquad (\tfrac{1}{2}\Delta)(a_2 + b_2x + c_2y)] \\ &= \frac{\mathrm{d}}{\mathrm{d}x}\left[\frac{1}{\Delta 2}(a_1a_2 + a_1b_2x + a_1c_2y + b_1a_2x + b_1b_2x^2 + b_1c_2xy + c_1a_2y + c_1b_2xy + c_1c_2y^2)\right] \end{aligned}$$

$$= \frac{1}{\Delta 2}(a_1 b_2 + b_1 a_2 + 2x b_1 y_2 + b_1 C_2 y + C_1 b_2 y)$$

$$= \frac{1}{\Delta 2}(a_1 b_2 + b_1 a_2 + 2 b_1 b_2 \bar{X} + b_1 C_2 \bar{y} + C_1 b_2 \bar{y}) \quad \text{(D-9b)}$$

where $\bar{X}$ and $\bar{Y}$ are centroidal coordinates of the element. Such averaging yields reasonably accurate results for small element sizes.

Application of Boundary Conditions

In a typical angiographic outline, let the regions C_m, C_a denote the mitral valve, aortic valve and the LV chamber wall outline, respectively. The nodal point i is recognized as the instantaneous location of the catheter tip that measures LV pressure at that point. The nodal points on C_m, C_a and C_w are designated by m, a, and w, respectively. The other nodal points in the LV chamber are designated by j. Thus, the total number of nodal points n is given by

$$n = 1 + j + m + a + w$$

Assembly of the element stiffness matrix for the whole LV chamber yields the system of equation

$$KU = Q \quad \text{(D-10)}$$

where U is a column vector containing $2n$ values of velocity components and $n/2$ values of pressure (if the pressure solutions are sought at $n/2$ number of points). Thus, the size of the assembled K matrix will be $(5n/2 \times 5n/2)$. Typically, in the assembled form of Eq. (D-10), the column vectors U and Q look like:

$$U = \begin{Bmatrix} u_n \\ v_n \\ p_{n/2} \end{Bmatrix}, Q = \begin{Bmatrix} R_n \\ R'_n \\ 0 \end{Bmatrix} \quad \text{(D-11)}$$

In order to solve this equation, suitable boundary conditions at nodes i, m, a, and w are introduced. We present below the detailed equations for such boundary conditions.

Boundary Conditions at Node i (catheter tip)

For node i, corresponding values of the set $p_{n/2}$ in the column matrix U (Eq. D-11) should be replaced by:

$$p_i = p_0 \quad \text{(D-12)}$$

where p_0 are recorded instantaneous LV catheter pressure. The other terms of the set p will give the pressure values at selected $(n/2 - i)$ points in the LV chamber, when Eq. (D-10) is solved. The corresponding terms for $p_{n/2}$ in the Q matrix are zero.

Boundary Conditions at Nodes w (LV chamber wall)

The velocity components u_w and v_w at nodes w are computed from the successive cineangiographic outlines of the LV chamber, as indicated in Paragraph III-A (Fig. 7).

Thus, for nodes w the corresponding values in the set u_n and v_n in the column matrix U are replaced by

$$u_w = \alpha_w \quad \text{(D-13)}$$
$$v_w = \beta_w$$

where α_w and β_w are the computed velocity components at chamber wall. It will be shown later that for such points (w) there is no need to compute the corresponding R_w and R'_w in the Q matrix.

Boundary Conditions at Nodes m and a (LV-valves interface)

Velocity boundary conditions at nodes m and a may be introduced in the same way as nodes w, i.e.,

$$\begin{aligned} u_m &= \alpha_m \\ y_m &= \beta_m \\ u_a &= \alpha_a \\ v_a &= \beta_a \end{aligned} \quad \text{(D-14)}$$

When the mitral valve or aortic valve is closed, the values of α_m, β_m, and α_a, β_a, respectively, are computed from the respective wall movement in the same way as α_w, β_w. However, when the valve is open, the velocity at the valve surface is computed by dividing volume changes (computed from successive angiograms) by the valve area. Obviously in such cases we are assuming a uniform velocity field at the valves during filling or ejection.

In Eq. (D-7), for non-boundary nodes, R_n and R'_n are zero; for special cases (e.g. at aortic valve) where, at some nodes, the pressure may be specified as p_a the solution for the corresponding velocity is sought, the corresponding R_a and R'_a term in the Q matrix are replaced by

$$R_a = \int_{C_a} N_a \bar{X}^* \mathrm{d}s$$
$$R'_a = \int_{C_a} N_a \bar{Y}^* \mathrm{d}s \quad \text{(D-15)}$$

For the assumption $\mu = 0$ at the open valve boundary,

$$R_a = -\int_{c_a} N_a N_x p_a \mathrm{d}s$$

$$R'_a = -\int_{c_a} N_a N_y p_a \mathrm{d}s \qquad \text{(D-16)}$$

Similar conditions may be applied to the mitral valve surface if the pressure at the mitral valve p_m is used as input.

When the velocities u_w and v_w are known in the U matrix, the corresponding values of R_w and R'_w in the Q matrix need not be computed. As an example, say

$$u_1 = \alpha_1, v_1 = \beta_1 \text{ is known} \qquad \text{(D-17)}$$

then

$$\begin{bmatrix} K_{11} & K_{12} & K_{13} & K_{14} \\ & K_{22} & K_{23} & K_{24} \\ & & K_{33} & K_{34} \\ & \text{Sym} & & K_{44} \end{bmatrix} \begin{bmatrix} u_1 \\ u_2 \\ v_1 \\ v_2 \end{bmatrix} = \begin{bmatrix} R_1 \\ R_2 \\ R'_1 \\ R'_2 \end{bmatrix} \qquad \text{(D-18)}$$

may be written as

$$\begin{vmatrix} 1 & 0 & 0 & 0 \\ & K_{22} & 0 & K_{24} \\ & & 1 & 0 \\ & \text{Sym} & & K_{44} \end{vmatrix} \begin{vmatrix} u_1 \\ u_2 \\ v_1 \\ v_2 \end{vmatrix} = \begin{vmatrix} \alpha_1 \\ R_2 - K_{21}\alpha_1 - K_{23}\beta_1 \\ \beta_1 \\ R_2 - K_{41}\alpha_1 - K_{43}\beta_1 \end{vmatrix} \qquad \text{(D-19)}$$

It may be noticed that R_1 and R'_1 need not be computed, and the above equations will automatically yield the solution:

$$u_1 = \alpha_1$$
$$v_1 = \beta_1 \qquad \text{(D-20)}$$

References

1. Brock R C: Functional obstruction of the left ventricle (acquired aortic subvalvular stenosis). Guy's Hospital Report 106, 221 (1957)
2. Cheng R T: Numerical solution of Navier-Stokes equation by the finite element method. Physics Fluids, 15 (1972)
3. Diamond G, Forrester J S: The effect of coronary artery disease and acute myocardial infarction on left ventricular compliance in man. Circulation 45, 11 (1972)
4. Gauer O H: Evidence in circulatory shock of an isometric phase of ventricular contraction following ejection. Fed Proc 9, 47 (1950)
5. Gauer O H, Henry J P: Negative ($-G_z$) acceleration in relation to arterial oxygen saturation, subendocardial hemorrhage and venous pressure in the forehead. Aerospace Med 35, 533 (1964)
6. Ghista D N: Rheological modelling of the intact left ventricle. In: Cardiac Mechanics (Mirsky I, Ghista D N, Sandler H, eds). Wiley-Interscience, New York 1974
7. Ghista D N, Sandler H: Elastic modulus of the human intact left ventricle – determination of and physiological interpretations. Med Bio Eng 13, 151–161 (1975a)
8. Ghista D N, Rao B N, Advani S: In vivo elastic modulus of the left ventricle: its determination by means of a left ventricular vibrational model, its physiological significance and clinical utility. Med Biol Eng 13, 162–170 (1975b)
9. Ghista D N, Hamid S: Finite element analysis of the human left ventricle. Comput Programs Biomed 7 (1977), 219–231 (1977)
10. Ingels N B, Daughters G T, Stinson E B, Alderman E L: Measurement of midwall myocardial dynamics in intact man by radiography of surgically implanted markers. Circulation, 52, 859–867 (1975)
11. Lin H S, Liu Y K, Ray G: System identificiation for material properties of the intervertebral joint. J Biomech 11, 1–14 (1978)
12. Mirsky I, Parmley W W: Evaluation of passive elastic stiffness for the left ventricle and isolated heart muscle. In: Cardiac Mechanics (Mirsky I, Ghista D N and Sandler H, eds) Wiley-Interscience, New York 1974
13. Nikravesh P E: Optimization in finite element analysis with special reference to three-dimensional left ventricular dynamics. Doctoral dissertation, Tulane University, New Orleans 1976
14. Ghista, D. N., Ray, G., Sandler, H.: Cardiac Assessment Mechanics. 1. Left Ventricular Mechanomyocardiography, a new approach to detection of diseased myocardial elements and states, Med. Biol. Eng. Comput., 18, 3, 241–280 (1980)
15. Ray G, Liu Y K, Chandran D B, Ghista D N, Sandler H: Potential energy distribution prior to ejection as a basis for post-surgical evaluation of left ventricular efficiency. Proc 29th ACEMB, 1976b, p 126
16. Ray G, Liu Y K, Chandran K B, Wieting D W: Pressure gradient changes in experimental acute myocardial infarction in dogs. Proc ASME Biomechanics Symposium, Yale University 1977

17. Ray G, Liu Y K, Chandran K B, Giles T D: Existence of a normal intraventricular pressure gradient: experimental validation in canine left ventricle. Circ Res (submitted) (1979)
18. Taylor C, Hood P: A numerical solution of the Navier-Stokes equation using the finite element technique. Comput Fluids Vol 1, No 1 (1973)
19. White R I, Criley J M, Lewis K B, Roos R S: Experimental production of intracavity pressure differences. Am J Cardiol 19, 221 (1967)
20. Yamada Y, Ito K, Yokouchi Y, Lamano T, Ohtsubo T: Finite element analysis of steady fluid and metal flow. Proc Int Symp on Finite Element Methods in Flow Problems, Univ of Wales, Wales 1974
21. Zienkiewicz O C: The Finite element in Engineering Science. McGraw-Hill, London 1971

Prof. Dhanjoo N. Ghista, Ph.D.
Head, Biomedical Engineering Dept.
Faculty of Medicine and Engineering
Chedoke-McMaster Rehabilitation Centre
Box 2000, Station 'A'
Hamilton, Ont. L8N 3Z5 (Canada)

Adv. cardiovasc. Phys., vol. 5 (Part III), pp. 84–103 (Karger, Basel 1983)

Transmission of Vibrations of the Heart to the Chestwall

J. Verburg
Laboratory of Medical Physics, Amsterdam, The Netherlands

Contents

Abstract

The transmission of heart sounds through the human body is not a clearly understood problem. Hitherto, its investigation, as reviewed in this chapter, has been phenomenological and qualitative, with no attempt made at physical formulation of transmission models. Herein the transmission of waves in a visco-elastic medium is presented. A short review is given on heart sound genesis theories. After which three transmission models are considered. Propagation along the surface is examined first. A second attempt is a pulsating source. Based on physiological data a third model, a mechanical dipole, is introduced and formulated.

The experimental results presented show a number of data which cannot be explained by propagation along the surface nor by the pulsating source. The dipole model however, is in agreement with all the experimental results and it is concluded that for the moment this model gives the best representation of the transmission of vibrations caused by motion of the heart valves. A firm quantitative calculation is however, not yet possible due to lack of reliable values of the material properties of the tissues between heart and skin.

I. Introduction

From early times the auscultating physician has been making use of the fact that vibrations originating in the heart are transmitted to the surface of the human body. Initially the "naked" ear was used to detect "sounds" from the body. Later, with the introduction of the stethoscope by the french physician Laennec in 1816 auscultation grew into an art, the result very much dependent on "that most important part of the total system that between the earpieces of the stethoscope" (Warren, cited by McKusick (1958)).

From a physical point of view, mechanical vibrations on the chestwall are converted to acoustic vibrations, which can be detected by human ears if the frequency and loudness lie within the range of the ear-sensitivity curve. With the introduction of mechano-electric transducers, it became possible to convert the mechanical vibrations into electric signals, which could be registrated easily and analysed next. The dutch physician Einthoven was one of the first to make registrations of heart vibrations, using a carbon microphone and a string galvanometer combined with photographic registration (Einthoven 1907). Phonocardiography was born.

The development of phonocardiography in the past years has essentially been directed to clinical use. Phonocardiography came to serve as an extra diagnostical tool besides auscultation and the electrocardiogram to detect abnormalities in the heart function. Many phonocardiographic methods have been developed. In the bibliography of Ravin and Frame (1971) many references can be found on clinical use and filter evaluation of phonocardiography. Since many investigators have been working on the optimisation of the technical aspects (such as microphones, slope of filter characteristics) for clinical use, the result has been a great variety of phonocardiographic equipment and often registrations made in one clinic cannot be read by a physician used to another apparatus in his own clinic (Groom 1970). As a result of this, phonocardiography has become in general a qualitative method.

A physicist, attempting to give a more quantitative base to phonocardiography, is confronted with the following problems:

1. What is the mode of transmission of vibrations, due to the heart action, to the surface of the chestwall?
2. Why is the maximum intensity of the sound, due to certain valves, not found directly over the site where the valves can be projected on the chestwall?
3. Is it possible to propose a model which gives a mathematical relation between the mechanical functioning of the heart and the vibrations detected on the surface of the body?

II. Earlier Investigations

The transmission of sound from the heart to the surface of the human body has been studied by a number of investigators. Feruglio (1962) studied the transmission to the chestwall of vibrations produced by a small turbo generator (a device very much like a siren, but for underwater applications) within the heart. The frequency of the noise produced was between 80 and 360 Hz. He found that sounds produced at the valves and at other places were transmitted well to certain places on the thorax but poorly, or not at all, to other places. Heintzen and Viets (1963) studied the transmission in the opposite direction by placing a vibration generator (a loudspeakersystem without cone) on the chestwall, while the transmitted signal was recorded in the heart by means of a phonocatheter. They employed frequencies between 50 and 1000 Hz. The transmission was markedly influenced by the cardiac cycle, a fact which is not so surprising, since the measuring site lies within the cardiac structures.

Zalter et al (1962) studied the transmission characteristics of the thorax of dogs by introducing sound, produced by an underwater loudspeaker, via a catheter into the cardiac chambers and recording on the bodysurface with air-microphones. Values for the attenuation of vibrations between 50 and 500 Hz are given. They find a maximum transmission in the vicinity of 120 Hz.

In the investigations of Faber and Burton (1962) and Faber (1963), the vibration sources are the heart valves. They studied the transmission of heart sounds to the body surface, more or less quantitatively. It was found that vibrations from the valves arrive first and are of maximum intensity, not at the places where one can project the sources (valves) on the surface, but on the classical auscultation areas. The mechanism of the transmission they propose is conduction along the heart structures and arteries through the body to secondary sources on the chestwall, from where they travel along the surface. They studied propagation by measuring the time differences between the arrival of vibration complexes at different sites. They find values of 1–15 ms depending on distance between the pick up sites. Studies of Shah (1963) using the same technique on patients with Starr-Edwards artificial valves contradict some of the findings of Faber. They find very small time differences and a different site of first arrival as compared with the findings of Faber.

So far no definite physical model of the transmission of heart sounds has been formulated. It is difficult to fit the investigations to each other and get a clear view of the nature of the transmission.

III. Models of Transmission

A. Transmission of Waves in Visco-Elastic Media

At first thought one automatically thinks of wave propagation as a means of transmission of vibrations. There is a similarity between geophysical vibrations and vibrations of the heart. In both cases there is a source and, at some distance on the surface, vibrations are recorded. However, a major difference is the short distance which exists between the source and the detection place in the human body as compared to the geophysical case. In geophysics there is transmission by compressional, shear and surface waves. One has to establish whether the wave concept can give an adequate description of the observed phenomena in and on the chest. Based on literature (Remington 1957, Fung 1972) one can, for the small displacements occurring during vibration transmission through the body, consider the tissues involved as linear visco-elastic.

Transmission of vibrations by waves in a viscoelastic medium will in principle not be different from that in normal elastic media. Both compressional and shear waves can exist and so can surface waves (a combination of the former two) along the boundary. Real material constants are exchanged for complex expressions in visco-elastic media. For the Lamé constants one can write:

$$\lambda = \lambda_1 + i\omega\lambda_2$$
$$\mu = \mu_1 + i\omega\mu_2 \qquad (1)$$

where in (1) λ_1 and μ_1 have the following relation to the technically employed material constants:

$$\mu_1 = \text{shear modulus of the medium} = G$$

$$\lambda_1 = \frac{2G \cdot v}{1 - 2v} \quad (v = \text{Poisson constant})$$

(2) λ_2 and μ_2 represent the bulk and shear coefficients of viscosity. (3) ω is the circular frequency involved and $i = \sqrt{-1}$.

In the same way as in normal elasticity, wave equations can be derived (Landau 1970). For monochromatic plane compressional and shear waves, the wave equations are:

$$\frac{d^2 u_l}{dx^2} + k^2 u_l = 0 \text{ and}$$

$$\frac{d^2 u_t}{dx^2} + h^2 u_t = 0 \qquad (2)$$

in which the wave numbers k and h are defined as:

$$k = \left(\frac{\rho\omega^2}{2\mu + \lambda}\right)^{1/2} \text{ and } h = \left(\frac{\rho\omega^2}{\mu}\right)^{1/2} \qquad (3)$$

where ρ is the density of the medium.

The viscous properties give two clear distinctions from the elastic case:
1. The propagation velocity is frequency dependent (dispersion).
2. The waves are damped.

For plane waves, the propagation velocities can be calculated from (Oestreicher 1951)

$$c_l = \omega/\text{RE}(k) \qquad c_t = \omega/\text{RE}(h) \qquad (4)$$

in which c_l and c_t are the velocities of propagation of compressional, respective shear waves (RE means the real part of). After some calculations one finds:

$$c_l = v_l \cdot \sqrt{2} \cdot \frac{(1 + K^2)^{1/2}}{\{(1 + K^2)^{1/2} + 1\}^{1/2}} \qquad (5)$$

and

$$c_t = v_t \cdot \sqrt{2} \cdot \frac{(1 + H^2)^{1/2}}{\{(1 + H^2)^{1/2} + 1\}^{1/2}} \qquad (6)$$

in which

$$v_l = \left(\frac{\lambda_1 + 2\mu_1}{\rho}\right)^{1/2}, \; v_t = \left(\frac{\mu_1}{\rho}\right)^{1/2},$$

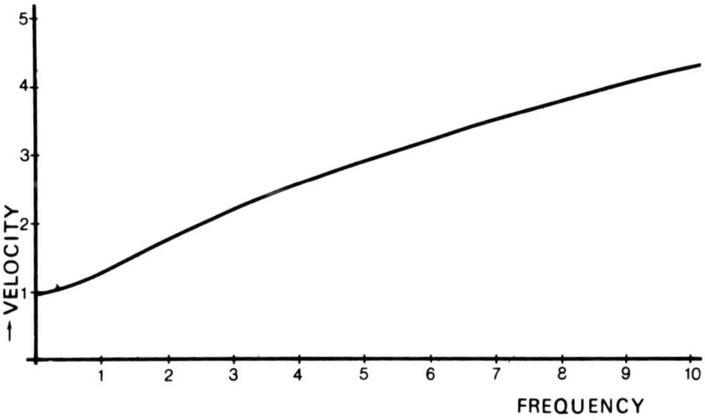

Fig. 1 The dimensionless propagation velocity of compression and shear waves $\frac{c_l}{v_l}\left(\frac{c_t}{v_t}\right)$ as a function of a dimensionless frequency $K(H)$.

K and H are dimensionless frequencies defined as:

$$K = \omega(\lambda_2 + 2\mu_2)/(\lambda_1 + 2\mu_1) \text{ and}$$
$$H = \omega\mu_2/\mu_1 \qquad (7)$$

The expressions for c_l and c_t have the same form. The curve of Fig. 1 gives the dimensionless propagation velocity $c_l/v_l(c_t/v_t)$ as a function of $K(H)$. It is clear that waves show dispersion due to the viscous properties. For the average values of the elastic properties of body tissues this dispersion is of no importance in the case of compressional waves in the frequency range met in phonocardiography. The shear waves however, show a definite dispersion in this frequency range. Dispersion causes a marked change in the form of a wave complex. The viscous properties are also responsible for damping of the waves. This means that the amplitude of the vibration will decrease, due to dissipative processes, faster than indicated by the spread through a volume.

The attenuation constant α can be calculated from the wave number (Oestreicher 1951). For plane compressional and shear waves one respective finds: α_l = imaginary part of k and α_t = imaginary part of h. Table 1 gives some values of velocities and attenuation constants for a tissue whose properties are assumed as follows:

$$\lambda_1 = 2.6 \times 10^9 \text{ N}\cdot\text{m}^{-2}, \lambda_2 = 0,$$
$$\mu_1 = 2.5 \times 10^3 \text{ N}\cdot\text{m}^{-2},$$
$$\mu_2 = 15 \text{ N}\cdot\text{s}\cdot\text{m}^{-2} \text{ and}$$
$$\rho = 1100 \text{ kg}\cdot\text{m}^{-3}.$$

B. Physiological, Anatomical and Physical Factors of Importance

1. Origin of Heart Sounds

The rather vague description one encounters when studying standard textbooks on the subject of genesis of heart sounds does give the reader the suspect feeling that in fact the precise knowledge of the phenomena is rather poor. Indeed as Craige (Craige 1976) in a recent editorial writes: "... the subject of the genesis of heart sounds is more controversial than ever."

Let us first see in short what happens during a contraction cycle of the heart and what can

Table 1 Values for velocity and attenuation of plane compressional and shear waves.

Frequency Hz	Phase velocity: Compression $m\cdot s^{-1}$	Phase velocity: Shear $m\cdot s^{-1}$	Attenuation constant: Compression m^{-1}	Attenuation constant: Shear m^{-1}
1	1540	1.5	1.8×10^{-10}	7.8×10^{-2}
10	1540	1.6	1.8×10^{-8}	7.8
100	1540	3.75	1.8×10^{-6}	1.3×10^{2}
1000	1540	13	1.8×10^{-4}	4.8×10^{2}

be detected by a phonocardiograph. Fig. 2 gives a sample of a signal obtained by an accelerometer glued to the chest wall. This signal can be regarded as a low frequency phonocardiogram. As in other normal phonocardiograms four complexes can be distinguished. The first vibration complex "the first sound" appears immediately after the QRS complex in the electrocardiogram. It coincides with the fast rising of the pressure in the ventricles, the closure of the atrio-ventricular valves, with the opening of the semi-lunar valves and the beginning of the ejection phase. In recordings which emphasize the high frequencies in the phonocardiogram this first complex is clearly split into two or more components.

At the end of the systole coinciding with the fast drop of the pressure in the ventricles another complex can be seen, "the second sound" very often in high frequency recordings clearly split of which in normal cases the first component coincides with the closing of the aortic valve and the second component with

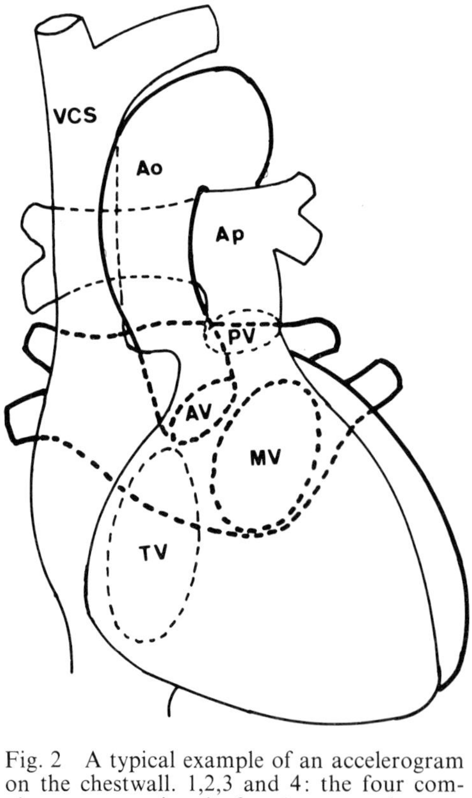

Fig. 2 A typical example of an accelerogram on the chestwall. 1,2,3 and 4: the four complexes representing the four heartsounds.

the closing of the pulmonary valve (Lewis 1962, Scher 1974).

Heart sounds have, due to their transient character, a continuous frequency spectrum. For the first sound the maximum intensity in this spectrum lies around 30 Hz, under normal circumstances, while for the second sound a maximum around 50 Hz is reported (Luisada 1965, McKusick 1958). Two other less clear complexes can be distinguished: the third sound, occurring in the rapid filling phase of the ventricles, and the fourth sound coinciding with the contraction of the atria. The genesis of these sounds nor the transmission to the chest wall will be dealt with.

During the last years a new discussion has started on the genesis of heart sounds. New methods such as ultra sound echo cardiography have given new data which made a reconsideration valuable.

Two major concepts are developed with respect to the first sound. The first is emerged from the classical theory which says that the high frequency vibrations are caused by valve closure or valve tension. Rushmer (Rushmer 1970) de-emphasizing the valvular cause, states as the origin of heart sounds: the sudden deceleration of the blood in the ventricle upon closure of the valves which sets the whole cardiohemic system into vibration. The splitting into two or more components of the first heart sound, was believed to be due to vibrations excited after closing of the mitral and tricuspid valve and possibly in the outflow tract of the left chamber upon the opening of the aortic valve. The second concept which attacks this view was developed by Luisada and co-workers (Luisada 1974). In their opinion the valves do not play a significant role in the production of vibrations and even more important they state that the whole first sound is completely a left heart phenomenon. The second component in the first heart sound is not as in the former concept attributed to tricuspid closure but to the ejection of blood into the aortic root. Evidence to this view is retracted from a time delay between the crossing of atrial and ventricular pressure, and the onset of the "mitral component" in animal experiments, and clinical observations during catherization procedures. In other animal experiments where the right heart was bypassed no change was noted in the duration or pattern of the first sound (Luisada 1974). Bypassing of the left heart, however, caused a disappearance of this sound. Based on studies of MacCanon (MacCanon 1969) it was concluded by calculations that valve tension can only account for some 10 percent of the total vibrational energy. Sakamoto et al (Sakamoto 1965) showed a relation between amplitude of the first heart sound and the pressure time derivative. On the other hand Waider and Craige (Waider and Craige 1975) found in their studies of subjects with clear splitted first heart sounds that the "mitral" component and "tricuspid" component occurred simultaneously with respectively mitral valve closure and tricuspid valve closure. The technique used in this case was ultra sound echo cardiography. In another study they demonstrated a high correlation between amplitude of the first sound and the separation of the valve leaflets. A contribution of ejection sounds to the first heart sound was in studies of Waider and Craige only demonstrated if deformations of the aortic valve were present.

Craige and co-workers (Craige 1976) see in these data a strong support of Rushmers postulates of the origin of the first heart sound with both left and right heart contributing. Although Luisada and co-workers also refer to Rushmers statements they accentuate the one sided origin and see no role of importance for the valves.

For the second sound there seems to be a general accepted explanation that blood masses, ventricular and vascular walls are set into vibration upon closing of the aortic and pulmonic valves. Time delays between aortic valve closure and aortic components in the second sound reported by Chandraratna (1975) question again the valve contribution but are strongly denied by other investigators (Rodbard and Mathews 1975, Craige 1976).

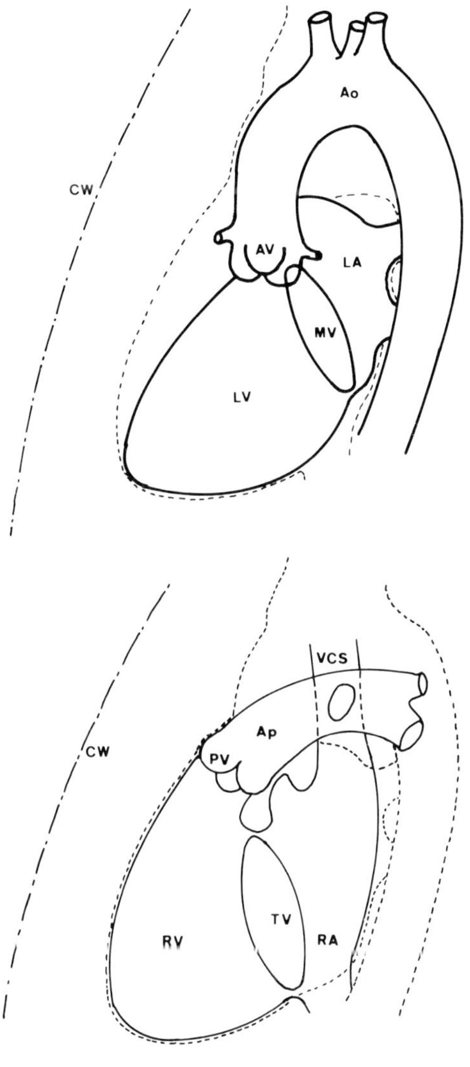

Fig. 3 The heart in the thorax, drawn in anterior view. VCS: vena cava superior, Ao: Aorta, Ap: arteria pulmonalis, PV: pulmonary valves, AV: aortic valves, MV: mitral valve, TV: tricuspid valve.

2. Position of the Valves

Although not a clear picture can be drawn on the real origins, one thing seems to be important for the transmission models:

The vibrations originate at the sites of the valves and are closely related to the velocity, de- and acceleration of the associated blood masses at the time of closure. It seems therefore also important for a transmission theory how valves and adjoining vessels are oriented in the chest. In Fig. 3 the heart, in the thorax, is drawn in an anterior view. The position of the valves is emphasized in this figure. In Fig. 4 (a and b), the left lateral views from the left and right heart are presented (Netter 1969). From the lateral views, especially, one can see that the planes through the aortic and pulmonary valves (across the corresponding arteries) intersect the chest wall at almost 90 degrees. The main direction of motion of these valves and the blood masses at closure is hence nearly parallel to the chest wall. On the other hand the main direction of motion of the mitral and

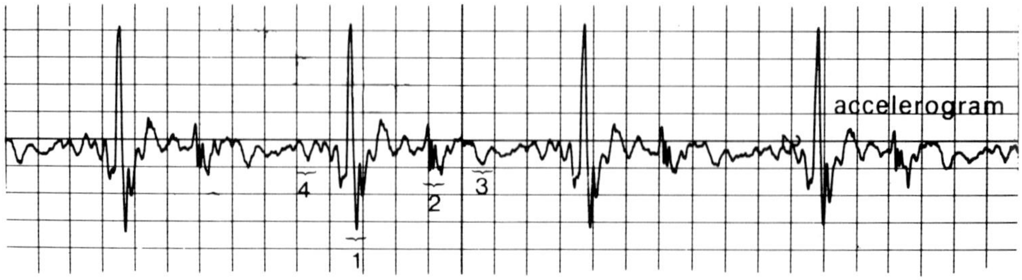

Fig. 4 a) Left lateral view of the left heart. LA: left atrium, LV: left ventricle, Ao: aorta, AV: aortic valves, MV: mitral valve, CW: chest wall. b) Left lateral view of the right heart. RA: right atrium, RV: right ventricle, PV: pulmonary valves, Ap: arteria pulmonalis, VCS: vena cava superior, TV: tricuspid valve, CW: chest wall. The stippled contour in both pictures is the heartshadow.

tricuspid valves and associated blood and tissue masses will be at an angle of about 45 to 60 degrees with the chest wall. Of course, during respiration the position of the valves will change. There will be a change in the distance to the chestwall and probably some rotation of the valve plane.

3. Mechanical Properties of the Tissues of the Chest

The vibrations discussed will travel to the chestwall surface through at least two sorts of tissue: lung tissue and the muscular connective tissue of the thoracic wall. The mechanical properties of these tissues are not very well known.

For the lung tissue, no direct measured values of Young's modulus and the shear modulus are to be found in literature. A density measurement is known from Dunn (1961) namely 400 $kg \cdot m^{-3}$ for lungs of dogs, having an estimated 2/3 of the original air left. Based on simple reasoning we estimate from volume and weight of lungs for the density of lung tissue a value between 200 and 300 $kg \cdot m^{-3}$. *Overall elastic properties* can be derived from models fitted to pressure volume curves, as is done by Mathews (1972) and Lambert (1973). However, the values they obtain are not completely comparable. Values for the shear modulus fairly agree; an average value in inspiration is 300 $N \cdot m^{-2}$, in expiration this decreases to 100 $N \cdot m^{-2}$. An important discrepancy exists for Poisson's ratio. Values between 0.2 and 0.5 are reported. A value of 0.5 Poisson's ratio means, given the relations between mechanical properties, a high value of Lamé's constant λ_1, thus giving a large compression wave velocity. Dunn (1961) reports a velocity of 650 $m \cdot s^{-1}$. Given the density and the shear modulus this leads to a value of Poisson's ratio of 0.5. This agrees very well with the suggestion of Mathews (1972) that in vivo lung tissue, considered as a continuum, behaves as an incompressable medium. Viscous properties of lung tissue are probably of great importance for the transmission, but no values are available at the moment. From values of the ultrasonic absorption given by Dunn (1961) the value of μ_2 is estimated at 300 $N \cdot s \cdot m^{-2}$.

Properties of the tissue closer to the surface are better investigated. The density of this tissue will not be much different from what is found for other soft tissues in the body and 1100 $kg \cdot m^{-3}$ is a good estimate. The velocity of compression waves in this tissues varies between 1500 and 1600 $m \cdot s^{-1}$ (Wells 1969) giving an estimate of 2.6×10^9 $N \cdot m^{-2}$ for the Lamé constant λ_1. The Lamé constants μ_1 and μ_2 can in general be calculated from mechanical impedance measurements on the chestwall, using the analysis of V. Gierke (1952). For soft tissues V. Gierke (1952) reports a rather wide spread of values. Mechanical impedance measurements of the chestwall have been done by several investigators. Unfortunately the values they find are not of the same order of magnitude but can differ a factor of almost 10 (Verburg and Vollenhoven 1979). We estimated the shear modulus μ_1 at 5000 $N \cdot m^{-2}$ and the viscosity coefficient μ_2 at 50 $N \cdot s \cdot m^{-2}$.

For the model calculations we adopted a homogeneous medium between the source (valves) and detection place (chestwall). From experiments we got the impression that the lung tissue is an important medium in the transmission path. Therefore, we made in most calculations for the mechanical properties the following compromise:

$\lambda_1 = 1.25 \times 10^8\ \mathrm{N \cdot m^{-2}}, \lambda_2 = 0,$

$\mu_1 = 300\ \mathrm{N \cdot m^{-2}}, \mu_2 = 50\ \mathrm{N \cdot s \cdot m^{-2}}$

and $\rho = 300\ \mathrm{kg \cdot m^{-3}}$

C. Propagation Along the Surface

As a result of the boundary conditions (normal and tangential forces are zero) there can exist waves along the surface of an elastic medium. These so called Rayleigh waves, in fact a combination of compressional and shear waves, have a velocity which is almost equal to the velocity of shear waves in the body. As is shown by Landau (1970), for a medium for which Poisson's constant is 0.5, the surface wave velocity is 0.96 times the shear wave velocity. A point on the surface describes an ellipsis when the wave is passing. And it is clear that horizontal as well as vertical displacements (accelerations) can be detected.

Franke (1951) showed that this also is true for visco-elastic media and that these surface waves show the same dispersion and damping as shear waves. The result of these data is that a complex wave front will change its form when propagated along the surface.

D. The Pulsation Model

We will now consider as a vibrating source a sphere whose radius will change sinusoidal in time, which we will call the pulsaton model. The vibrational field of such a source can be described by compression waves. The associated wave equation, in spherical coordinates, is given by:

$$\frac{1}{r^2}\left(\frac{\mathrm{d}}{\mathrm{d}r}\left(r^2\frac{\mathrm{d}s}{\mathrm{d}r}\right)\right) + k^2 s = 0 \qquad (8)$$

where $r(= \sqrt{x^2 + y^2 + z^2})$ is the distance between any point and the origin, s is the radial displacement and k defined as in section III A. If we assume an infinite medium, no reflected wave can be expected. If the radius of the sphere is given by a and if the centre of the sphere coincides with the origin we can calculate from Eq. (8) and the boundary condition

$$s = s_0 e^{i\omega t} \qquad \text{for } r = a,$$

that the radial displacement s is given by:

$$s = s_0 \cdot \frac{a^2}{r^2} \cdot \frac{e^{ika}}{e^{ikr}} \cdot \frac{1 + ikr}{1 + ika} e^{i\omega t} \qquad (9)$$

In view of the maximum in the frequency spectrum of valve vibrations we calculated, at a frequency of 30 Hz, the relative amplitude of the vibration as a function of the distance to the centre of the sphere (of radius = 1 cm). The result is presented in Fig. 5. If we wish to apply this model, the first thing to note is that the actual medium is finite; it is precisely at the boundary that we need to calculate the displacements. In general there will be reflections which influence the amplitudes at the surface. If the distance from the source to the surface is at least 3 to 4 times bigger than the radius of the source, it can be concluded that it is plausible to assume that no multiple reflection will take place, due to rapid attenuations. For displacements perpendicular to the surface, the amplitudes in that case will be at most a factor

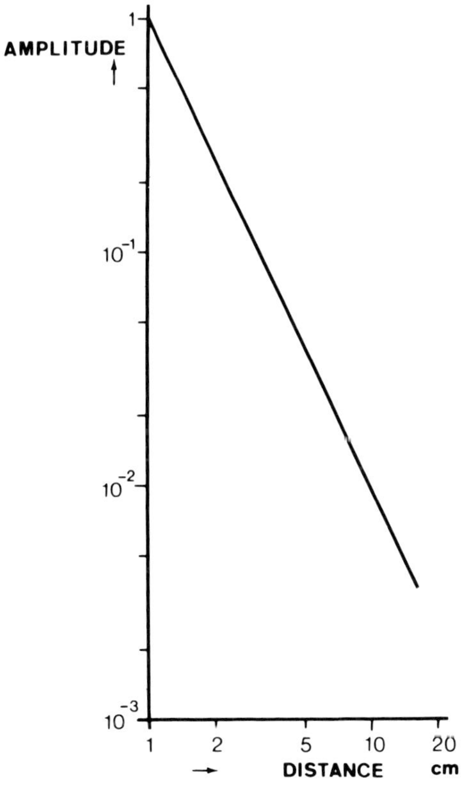

Fig. 5 Amplitude in radial direction, in the field of a pulsating sphere as a function of the distance to the centre of the sphere. Frequency 30 Hz. $\lambda_1 = 1.25 \times 10^8\ \mathrm{N \cdot m^{-2}}, \lambda_2 = 0, \mu = 300\ \mathrm{N \cdot m^{-2}}, \mu_2 = 50\ \mathrm{N \cdot s \cdot m^{-2}}, \rho = 300\ \mathrm{kg \cdot m^{-3}}$.

2 greater, than if one calculates the amplitude at this place, assuming an infinite medium.

For two entirely different values of λ_1, we have calculated under these conditions the relative amplitude and phase of vibrations perpendicular to a plane at 5×10^{-2} from the centre of a spherical source with a radius of 1×10^{-2}, using the formula 9 Fig. 6 gives results of these calculations. Phase shifts between displacements perpendicular to the plane can be interpreted as delays caused by a wavefront travelling along this plane. It is clear that with a big value of λ_1 this wave velocity is very large. With the low value of λ_1, which is unlikely to exist as we have seen in section III B 3, one finds an apparent velocity of $5.3\ \mathrm{m \cdot s^{-1}}$. Furthermore, it can be concluded from formula (9), that the solution is quite symmetrical round the projection of the centre of the source on the surface.

E. The Dipole Model

As already pointed out in section III B 2, the components of the heart valves and surrounding tissue can be expected to move perpendicular to the plane across the valve gap. Vibrations, transferred to the surrounding tissues, will probably have the same direction. It seems therefore reasonable to examine a model in which the source has mainly a vibration along one axis. As an approximation a spherical source vibrating along one axis in an infinite visco-elastic medium is chosen. The field of such a source has been calculated by Oestreicher (1950). He finds for the displacement vector $s(s_x, s_y, s_z)$ in a point x, y, z with radius $r(= \sqrt{x^2 + y^2 + z^2})$:

$$s = \left[-A\,\mathrm{grad}\left(\frac{h_1(kr)}{kr}x\right) + B\left\{2h_0(hr)\cdot\mathrm{grad}(x) - h_2(hr)\cdot r^3\cdot\mathrm{grad}\left(\frac{x}{r^3}\right)\right\}\right] s_0 e^{i\omega t} \tag{10}$$

in which h_0, h_1 en h_2 are spherical Hankelfunctions. (Spherical Hankelfunctions are defined as $h_n = j_n - i\,n_n$, j_n and n_n being spherical Besselfunctions of the first and second kind of order n (Morse 1948, Abramowitz 1968).) Making use of some relations between Hankelfunctions (Abramowitz 1968) we calculate for the components of s

$$s_x = -\frac{A}{k^3 r^3}\left[(i - kr) - \frac{x^2}{r^2}(3i - ik^2r^2 - 3kr)\right]\exp i\omega\left(t - \frac{k}{\omega}r\right) + \frac{3B}{h^3 r^3}\left[(-i + ih^2r^2 + hr) + \frac{x^2}{r^2}(3i - ih^2r^2 - 3hr)\right] \exp i\omega\left(t - \frac{h}{\omega}r\right) \tag{11}$$

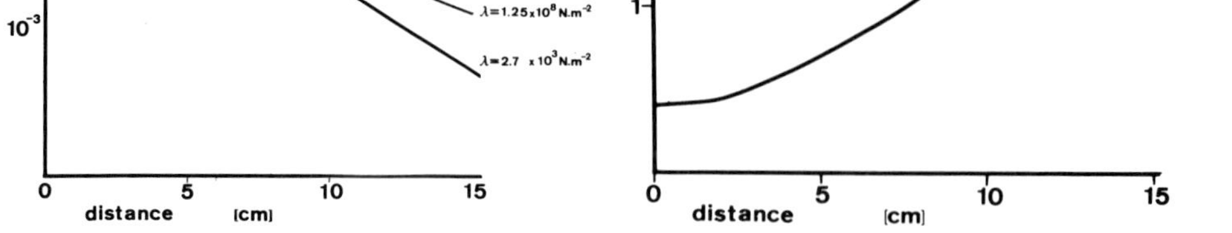

Fig. 6 Amplitude and phase of vibrations perpendicular to a plane, at a distance of 5 cm from the centre of a pulsating sphere, as a function of the distance along the plane from the projection of the center of the source on the surface. Calculations are made for two values of the Lamé constant λ_1. Frequency 30 Hz. $\lambda_2 = 0$, $\mu_1 = 300\ \mathrm{N \cdot m^{-2}}$, $\mu_2 = 50\ \mathrm{N \cdot s \cdot m^{-2}}$, $\rho = 300\ \mathrm{kg \cdot m^{-3}}$. N.B. The phase is so small in the case of a large value of λ_1, that it cannot be drawn in the figure.

and

$$s_y = -\frac{A}{k^3r^3}\left[-\frac{xy}{r^2}(3i - ik^2r^2 - 3kr)\right]$$
$$\exp i\omega\left(t - \frac{k}{\omega}r\right)$$
$$+ \frac{3B}{h^3r^3}\left[\frac{xy}{r^2}(3i - ih^2r^2 - 3hr)\right]$$
$$\exp i\omega\left(t - \frac{h}{\omega}r\right) \quad (12)$$

with an identical expression for s_z as for s_y, wherein y is replaced by z. Here in h and k are defined as in section III A. From the boundary conditions $s_z = s_y = 0$ and $s_x = s_0e^{i\omega t}$ when $r = a$,
it follows that

$$A = -i\frac{(3 + 3iha - h^2a^2)k^3a^3\exp(ika)}{k^2a^2(1 + iha) + (2 + 2ika - k^2a^2)h^2a^2}$$

$$B = i\frac{(3 + 3ika - k^2a^2)h^3a^3\exp(iha)}{3[k^2a^2(1 + iha) + (2 + 2ika - k^2a^2)h^2a^2]} \quad (13)$$

with a = radius of the vibrating sphere
s_0 = amplitude of the vibrating sphere

From Eqs. (11) and (12) it can be seen that displacements in the field can be described as the sum of two kinds of waves with different phase velocities and different damping factors. Each of these terms is multiplied by a different factor, determined by geometry. Comparison with the expressions in III A, shows that one can speak about a shear and a compression part of the displacement. If specific values for the material constants are inserted in formulas (11) through (13) it appears from calculations that in the vicinity of the sphere (1 to 5 times the radius of the sphere) both contributions are of the same order of magnitude (Fig. 7). The absolute values of these contributions close to the sphere are much bigger than the displacement itself; thus the presentation of the displacement with waves is, at that place, very formal. At a distance greater than 10 to 12 times the radius of sphere, the contribution of the compressional waves very soon completely dominates in the displacement. The location of this transition is highly dependent on the frequency. At high frequencies the transition takes place closer to the sphere.

When this model is employed for vibrations of valves, one has to consider the fact that the medium involved is not infinite. It follows from the calculations that the amplitude of the

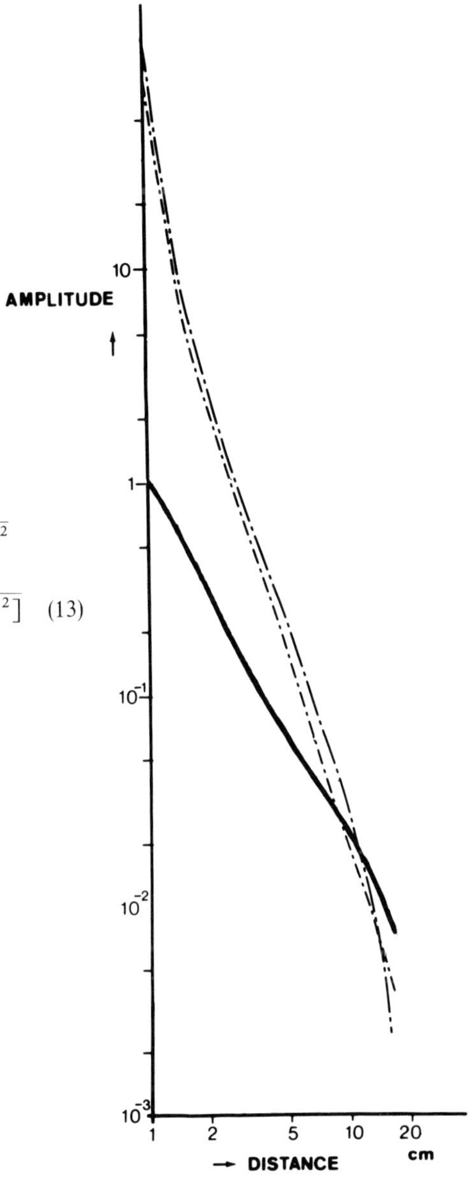

Fig. 7 Amplitude of vibrations in the field of an oscillating sphere, in an infinite visco-elastic medium (dipole model), as a function of the distance to the centre of the source, in a direction which has an angle of 45 degrees with the vibration direction. Frequency 30 Hz. $\lambda_1 = 1.25 \times 10^8\ N \cdot m^{-2}, \lambda_2 = 0, \mu_1 = 300\ N \cdot m^{-2}, \mu_2 = 50\ N \cdot s \cdot m^{-2}, \rho = 300\ kg \cdot m^{-3}$. —·—· — shear wave part of the vibrations. —·—·—·—·—·— compression wave part. ——— combined result.

vibrations is diminishing very fast. Multiple reflections are therefore very unlikely and the error in calculating the vibrations perpendicular to the surface, assuming an infinite medium, will be small for the same reasons as in the case of the pulsating sphere.

Under these conditions we have calculated, for two values of the Lamé constant λ_1, the amplitude and phase of vibrations perpendicular to the observation plane as a function of the distance to the projection on the observation plane of the centre of the source. The angle between the vibration direction of the sphere and the plane is chosen as a parameter. To compare the results with those of the pulsation model, we also adopt for this model the data: radius of the source (a) = 1×10^{-2} m, distance to the observation plane = 5×10^{-2} m, frequency 30 Hz. Fig. 8 gives results. A notable feature is that a large variation in λ_1 surprisingly gives no major changes. This means that one can expect, that variation of the mechanical properties between subjects, will not appreciably affect the results.

Another feature that attracts attention is the very fast changing phase shift at small angles in the vicinity of the projection of the source on the surface (Fig. 8b, d) change of about π radians, so in effect a phase reversal occurs. One might expect a change in sign of vibrations over a small distance if vibration direction of the source makes a small angle with the observation plane.

Thirdly, big changes appear in the phase characteristics when the angle between vibration direction and the observation plane is changed (Fig. 8d). As with the pulsation model the phase difference between two places on the plane can be interpreted as caused by wave propagation along the plane and the velocities of these "waves" are very much dependent on the angle between the vibration direction and the plane.

From the amplitude curves (Fig. 8a, c) it is

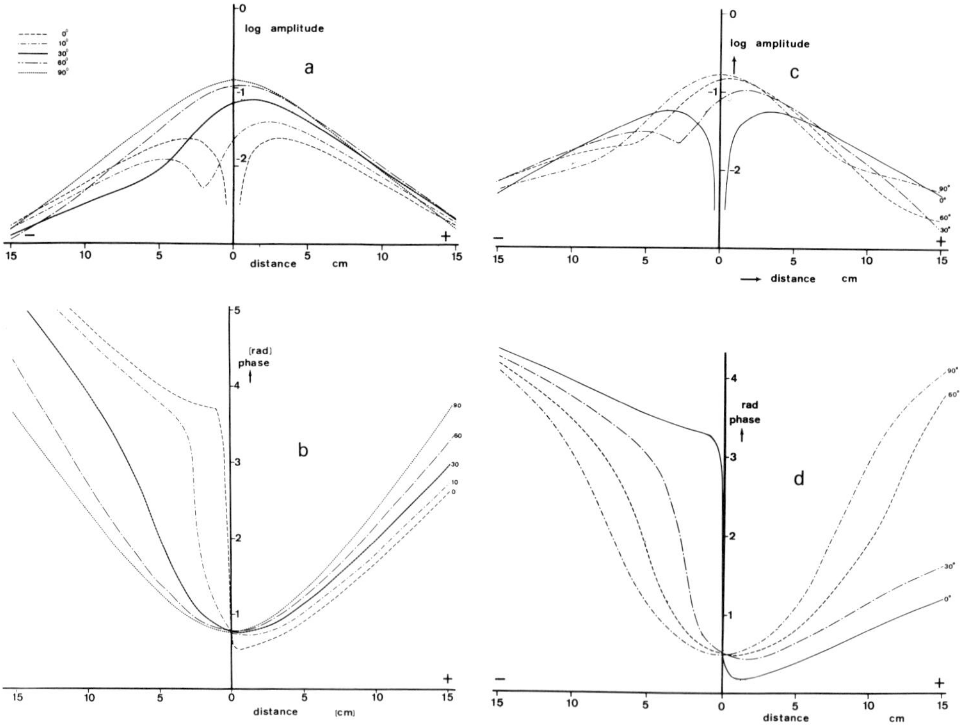

Fig. 8 Amplitude and phase of vibrations perpendicular to a plane, at a distance of 5 cm from the center of an oscillating sphere (dipole model) as a function of the distance along the plane from the projection of the center of the source. $\lambda_2 = 0$, $\mu_1 = 300\ \mathrm{N \cdot m^{-2}}$, $\mu_2 = 50\ \mathrm{N \cdot s \cdot m^{-2}}$, $\rho = 300\ \mathrm{kg \cdot m^{-3}}$. Radius of sphere: 0.01 m, frequency of the vibration 30 Hz. In a) and b) the value of λ_1 is $2.7 \times 10^3\ \mathrm{N \cdot m^{-2}}$, in c) and d) $1.25 \times 10^8\ \mathrm{N \cdot m^{-2}}$.

further clear that the maxima, except in the case of an angle of 90 degrees between the vibration direction and the surface, do not exactly lie on the place where the source can be projected on the observation plane. Calculations with other frequencies show the same effects of the phase reversal and the shift of the maxima in the amplitude curves (Verburg 1975). If the apparent velocity along the plane is calculated from phase characteristics like Fig. 8b and d (velocity = circular frequency/gradient of phase characteristics) we find the results of Table 2.

Table 2 Calculated apparent velocities

frequency	apparent velocity in $m \cdot s^{-1}$	
Hz	60°	30°
20	3.5	10
30	4.7	18
50	6.2	22
100	8.7	38

Inspiration and expiration show two effects:
1. change of the distance between valves and the chestwall and
2. changes in the angle between the motion direction of the valve and the chestwall.

It is further probable that the elastic properties of the medium between heart and chestwall change during respiration. As can be seen from Fig. 8a change in direction of the vibration of the source can give a change in apparent velocity since the gradient of the phase characteristic changes for other angles. A change in distance from the source to the surface can also be introduced in the model. Especially when the angle between the vibration direction and the surface is large (60 à 90 degrees) calculations give a marked change in apparent velocity (Verburg 1975). When the distance from the source to the surface changes during respiration from 5 cm to 4 cm, in the model a change in apparent velocity from 19 to 9 $m \cdot s^{-1}$ is calculated. During respiration the mechanical properties of lung tissue are changing (section III B 3). Introduction of this change in the model, however, show almost no changes in the apparent velocity (Verburg 1975).

We simulated this dipole model on a computer. As input signal of the dipole source a single, biphasic excursion is chosen, which is drawn in the top left corner of Fig. 9. The reason for chosing such a signal lies in the single character of valve vibrations. In the rest of the figure the amplitudes of vibrations perpendicular to the surface are given at various distances from the projection of the centre of the source on the surface and for varying values of the angle between the direction of vibration and the surface. The radius of the source is 1 cm, the distance to the surface is 5 cm. Two important features can be seen from this figure: The sudden change in sign of the signal at the observation plane at small angles, when the distance changes only a small amount. Secondly the location of the maximum of the signal on the plane, which is not coinciding with the projection from the centre of the source on the plane of observation. In section IV C an evaluation of the models will be made based on the experimental results of section IV B.

IV. Experimental Investigations

A. Methods

To investigate the vibrations of the chestwall, use is made of small accelerometers. These transducers have the following advantages:
1. The mechanical impedance of the transducer is small compared with that of the chestwall at the place of contact. The small weight causes no mentionable pressure to the chestwall. So the mechanical impedance of the chestwall is not influenced by it, which results in a better transfer of vibrations over a frequency range from 0–500 Hz (Verburg 1975). Fig. 10 gives values of the mechanical impedances of the chestwall as well as of the accelerometer and the mentioned transfer, expressed in a factor A (adaptation), all as functions of the frequency.
2. They are easy to calibrate. In the construction of the accelerometers, use is made of piezo-resistive materials (Verburg and Strackee 1974) giving low internal resistance (less hum) and a better low frequency response as compared to piezo-electric devices.

A number of these accelerometers (varying from 1–12) are glued to the chestwall with a double adhesive tape. They are connected to amplifiers through thin flexible cables. In all experiments an ECG signal was recorded – in general lead II according to Einthoven. After

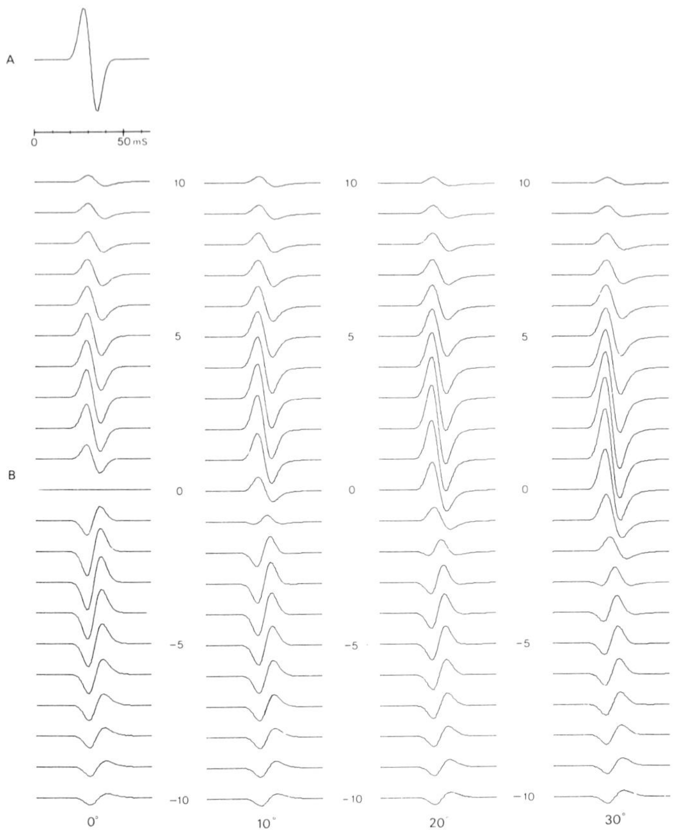

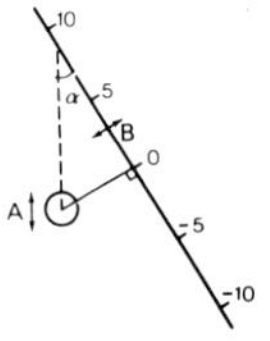

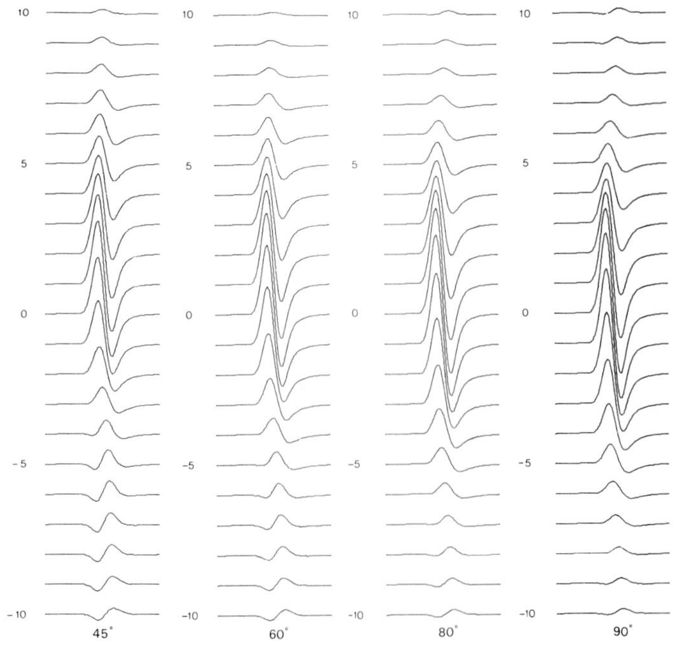

Fig. 9 Simulation with the dipole model. In the top left corner is shown the simulated valve vibration; in the right top a drawing of the simulated situation is shown. Under *B* the responses perpendicular to the plane are given for some sites on both sides of the projection of the centre of the source and for several angles between vibration direction and the surface. Vertical scale under *B* is bigger than that of *A* by a factor of 10.

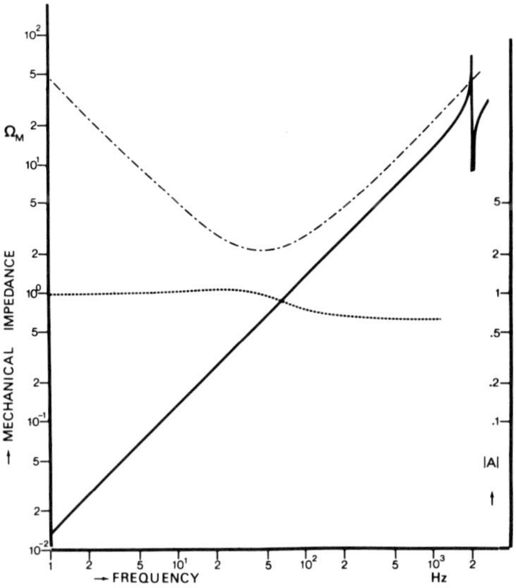

Fig. 10 Mechanical impedance of the chest wall (—·—·—·—·—) (surface area 4.9 cm²). Mechanical impedance of a small accelerometer (————) (contact surface 4.9 cm²) and adaptation *A* of this accelerometer to the chest wall (..........) as a function of the frequency. Impedance is given in Ω_m which is equivalent to $N \cdot s \cdot m^{-1}$.

amplification the signals can be observed on an oscilloscope and registrated on a paper recorder (ELEMA). The signals meant for further calculations are recorded on an F.M. taperecorder (AMPEX FR 1300) with a speed of 30 IPS. Calculations on the signals are made on a general purpose computer (DIGITAL PDP-9). Display of the data can be performed on a digital plotter (Calcomplot). A block diagram of the experimental set up is given in Fig. 11.

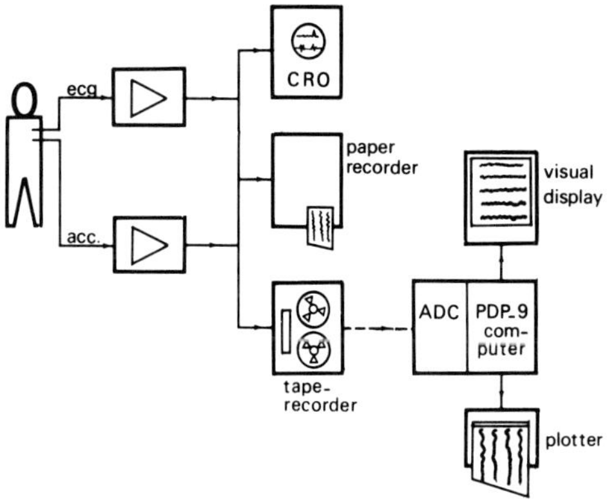

Fig. 11 Block diagram of the experimental set up for recording and operating of acceleration signals.

B. Results

Fig. 12 gives the typical results of a measurement of the accelerations at 12 places on the chestwall of a male subject (38 years old, expiration in supine position). The positions of the accelerometers lie on a line, mid-claviculair left side. Sensitivities of all the channels are the same. The upper track in the picture is a part of an electrocardiogram. Track nine gives the

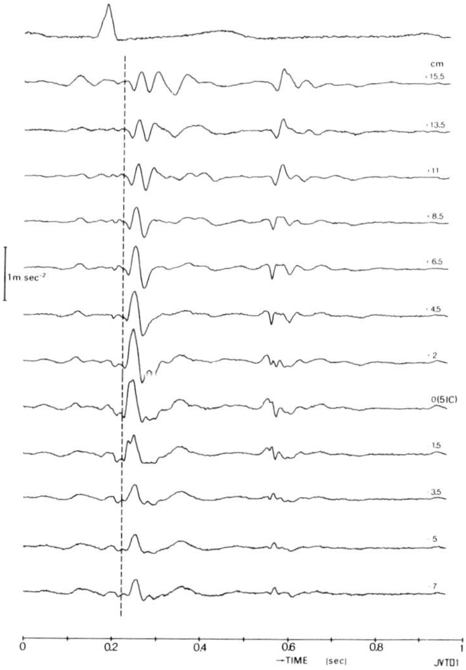

Fig. 12 Accelerogram recordings on the chest wall. Accelerometers are placed on a line parallel to the sternum, midclavicular left side. Distance to the accelerometer in the 5th intercostal space is given on the right side in the figure. Positive deflection means an outward direction of the acceleration. Measurements are made during expiration.

signal in the 5th intercostal space. The positions of the others are given with respect to this one. Positive deflection implies an outward acceleration. It is clear from the figure that time shifts exist between the acceleration signals at various sites. These time shifts are different for the first and second sounds. The difficulty is how to establish this time shift, whether to use the tops, valleys or zero-crossings in the signals as reference points. We used two techniques: first a straight forward measurement of time shifts between tops or valleys with a ruler. The other technique makes use of cross-correlation methods. The results of two techniques fairly agree (Verburg 1975). The repetition of the same experiment in inspiration gives other time shifts. When apparent velocities are calculated from the time shifts and the position differences Fig. 13 by a two sided linear regression (Verburg 1975) one finds a distinct difference between these velocities in inspiration and expiration, with several subjects (Table 3).

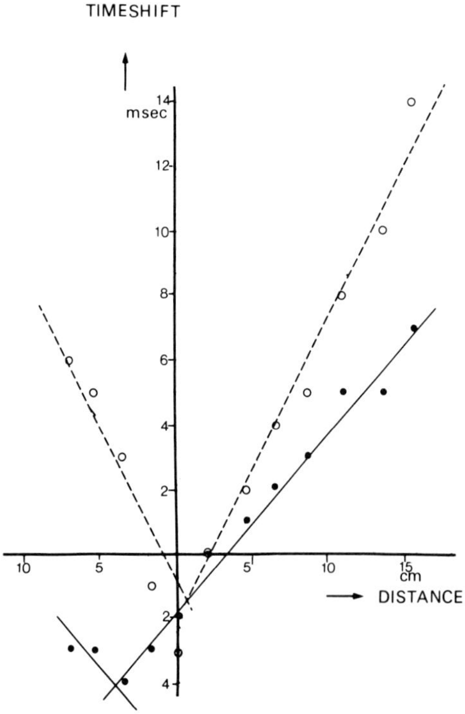

Fig. 13 Time shifts of a first tone complex in accelerograms as a function of the distance along a line on the chest wall, left midclavicular, in inspiration and expiration. The lines through the points have been obtained by a two sided linear regression.

Table 3 Experimentally found apparent velocities (1st sound)

subject	apparent velocity in $m \cdot s^{-1}$		
	expiration	inspiration	method
J.V.	10.5	18	correlation
P.B.	11.4	15.9	ruler
	8.5	15.6	ruler
	10.8	19.6	correlation
	9.5	16.7	correlation
F.O.	11.4	14.7	ruler
	12.6	14.0	ruler
G.L.	8.2	14.0	ruler
	8.4	17.2	ruler

The values of the velocities thus found are comparable with those mentioned by Faber and Burton (1962), they fairly agree with those calculated given in Table 2.

In some preliminary experiments we measured time shifts over a large part of the chestwall and, by filtering of the signal, for several frequencies. We are able to draw charts of isochrones (lines with equal time shifts in respect to one reference signal) for these frequencies. In Fig. 14 typical examples of such charts are given. From these charts it can be concluded that no circular symmetry exists for the "spread" of the vibrations (A similar conclusion can be drawn from the experiments of Hök et al (1978)) and that the apparent velocity increases with increasing frequency.

In many of registrations made on normal subjects on a left midclavicular line one sees in the second sound complex an almost complete reversal of the signal (Fig. 15). In most cases the channel between the two reversed channels shows a minimum amplitude. We will discuss these findings in section IV C.

To establish whether surface waves play a role in the transmission of the sounds we examined if, as one can expect, accelerations exist parallel to the surface. We therefore simultaneously measured acceleration perpendicular and parallel to the surface. In Fig. 16a the direction of the sensitivity of the accelerometer parallel to the surface makes an angle of 90 degrees with the sternum. In Fig. 16b this direction is parallel to the sternum. It is clear from both figures that accelerations exist parallel to the surface. This could be the effect

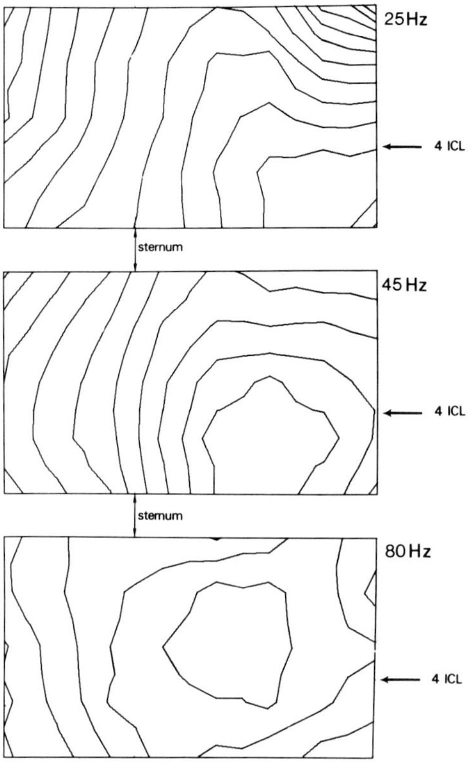

Fig. 14 Isochrone charts: charts with lines of equal time shifts on the chestwall for 3 frequencies. The time shifts between two lines is 2 ms.

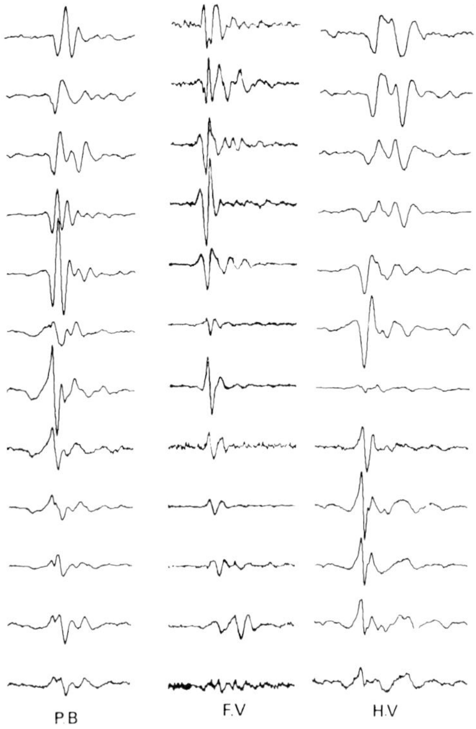

Fig. 15 Accelerograms recordings on the chest wall showing second sound complexes, in which a sudden reversal can be seen.

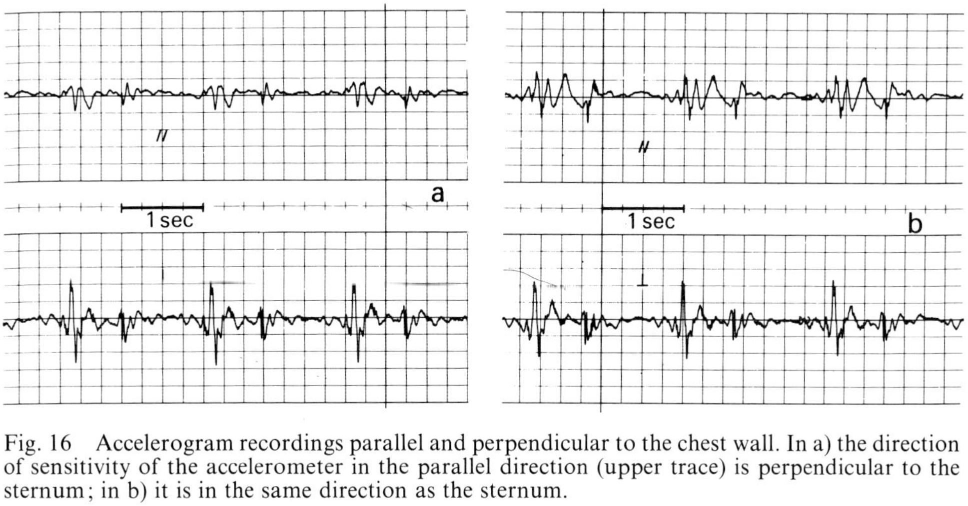

Fig. 16 Accelerogram recordings parallel and perpendicular to the chest wall. In a) the direction of sensitivity of the accelerometer in the parallel direction (upper trace) is perpendicular to the sternum; in b) it is in the same direction as the sternum.

of a passing surface wave. Ballistic effects, due to the gross heart action, may however, also play a role.

If it is assumed that surface waves play a role in the transmission, it should expected that the placing of a screen around an accelerometer, in close contact with the chestwall, will influence the wave pattern. Figs. 17a and b give registrations of an experiment to investigate these effects. Fig. 17a gives signals of accelerometers with a sensitivity perpendicular to the surface under normal conditions. Fig. 17b gives the accelerograms from the same places, but now the lower trace is from a screened accelerometer. We see no major differences between Fig. 17a and Fig. 17b. In section IV C we will compare the results obtained, with the predictions of the models.

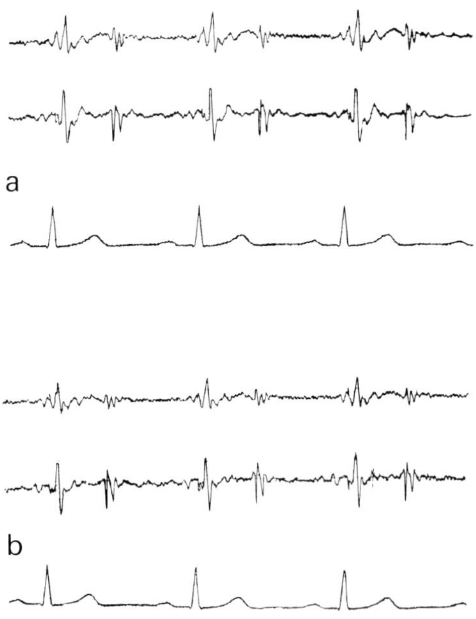

Fig. 17 Accelerogram recordings on the chest wall.
a) at two places without screening
b) at the same places as in a), but a screen is placed around the accelerometer giving the first track.
The third track in both the pictures is the ECG signal.

C. Comparison of Experimental Data with Those Provided by the Models

The experimentally found apparent velocity along the surface has a value in the order of 10 $m \cdot s^{-1}$. The value of surface wave velocity is (section III C) 0.96 times the shear wave velocity. Using the fact that we estimated the Lamé constants μ_1 and μ_2 for chestwall tissue at higher values than those of Table 1 (section III B 3) one can predict a value of around 10 $m \cdot s^{-1}$ for the propagation velocity along the surface at a frequency of 30 Hz. This is in good agreement with the mentioned experimental value. The dipole model gives values (Table 2) which are of the same order of magnitude and do not contradict the experimental value. The value obtained with the pulsation model is only comparable when a low value of λ_1 is assumed. As mentioned in section III B 3 a large value of λ_1 seems plausible for both the chestwall tissue and the lungtissue. Thus the experiment contradicts the data provided by the pulsation model.

The different values of the apparent velocity in inspiration and expiration do not agree with the assumption of surface waves, since in normal inspiration and expiration the path length is practically equal. Assuming the dipole model it can be explained by the change in distance from the valves to the surface and/or change in direction of the valve plane.

The charts of isochrones (Fig. 14) show no circle symmetry nor a first arrival at the site where the source can be projected on the surface and in this respect contradicts the pulsation model. It does not give difficulties with the dipole model. With propagation along the surface it can be explained if an inhomogenity of the chest surface is assumed, which seems plausible in view of the results of Faber (1963). The isochrone charts show that with increasing frequencies the apparent velocity along the surface also will increase. This does not agree with the calculations of the pulsation model. As the velocity of surface waves has the same frequency dependance as the shear waves (Fig. 1) it does not contradict the propagation along the surface. It also agrees with the dipole model analysis (Table 2).

That the maximum intensity of heart sounds lies not at the projection sites of the heart valves, as found in clinical practice, cannot be explained by the pulsation model, since it has a complete symmetrical solution. The dipole model does provide with a maximum beside the projection (Fig. 8a, c and Fig. 9). In the propagation along the surface it is an a priori.

All three models do predict acceleration along and perpendicular to the surface, in agreement with the experiment. Influence of screens round accelerometers is negligable, a fact which is not compatable with propagation along the surface. The reversal of sign indicated by the phase characteristics of the dipole model and shown in the simulation of this model (Fig. 9), is found in the experimental recordings of the second sound (Fig. 15). As the main contributions to this sound is coming from aortic and pulmonic valves, having a small angle between the vibration direction and the chestwall (Fig. 3), these findings agree very well with the data provided by the dipole model (Fig. 9). It cannot be explained by the pulsation model nor by propagation along the surface.

V. Discussion

In the forgoing sections it was assumed that the medium involved was homogeneous and isotropic. It is evident that these assumptions are not strictly representative of the chest tissues. Introducing an inhomogeneous and/or anisotropic medium will complicate the models without giving more significant information.

The assumption of an incompressible lung tissue may seem a bit confusing, but it should be kept in mind that the term incompressible makes sense only when talking about infinitely small displacements, where a linear relation exists between deformation and load. This certainly is not the case when respiratory displacements are considered, however for the small deformations and for the frequencies related to valve vibrations it seems reasonable to assume an incompressible lung tissue.

The comparison of experimentally found data provided by the models as summarized in Table 4 shows that the pulsation model nor propagation along the surface can explain all experiments.

The dipole model is in agreement or does not contradict any of the experiments done. That Shah et al (1962) could not measure on patients with Starr-Edwards artificial valves time shifts of the same order as Faber (1962) did, can also be explained by the dipole model.

Table 4 Comparison of the experiments with the models

Experimental data	Pulsation model	Dipole model	Propagation along surface
Value apparent velocity	–	+	+
Change of apparent velocity (inspiration and expiration)	–	+	–
Isochronal charts	–	+	+
Shift of maximum intensity of sounds	–	+	+
Accelerations parallel and perpendicular to the surface	+	+	+
Influence of screen	+	+	–
Sign reversal of signal	–	+	–

+ no contradiction with model
– contradiction with model

Vibrations of artificial valves have in general a frequency spectrum which has a maximum at much higher frequency (> 200 Hz) than normal values. The dipole model should hence be applied at a much higher frequency. Calculations show, however (Verburg 1975) that at high frequencies (> 200 Hz) phase differences between vibrations or different places on the surface are very small, independent of the angle between motion direction and surface. Thus it can be expected that the time shifts in practice will be very small.

If we consider all these facts we conclude that, for the present, use of the dipole model provides a very good description of the transmission of heartsounds. It can explain the clinically well known fact that heart sounds do not have their maximum intensity over the site where the valves can be projected on the chestwall.

The dipole model gives a relation between internally produced and externally registrated vibrations, and thus gives a solution of the "forward" problem in phonocardiography. In principle, one could, based on this model, give a quantitative description of internal vibrations based on vibrations measured at the surface, the "backward" problem. In this way the dipole model has been adopted by other investigators (Kozmann 1977, Cheung 1977)

A better knowledge of the mechanical properties of the tissues involved and of dimensions within the thorax will be necessary to get reliable results. Another necessity is to adopt an analysis in a finite visco-elastic medium. As the order of the differential equations is four in the elastic case and since the boundary is very complex, analytic solutions are probably not easy to find or even impossible. Numerical integration using digital computers will be necessary to solve the problem.

Another point of discussion can be what size one should attach to the source. We have adopted a size with approximately the dimensions of the valve orifices, which decision is made on a slight preference to postulate the source near the valves. If the concept of the heart sounds origins forwarded by Luisada and co-workers proves true a larger source with dimensions comparable with those of the left ventricle should be more appropriate. Model calculations show however, that qualitatively no different results are obtained and that the quantitative differences are small.

A final conclusion from the studies presented here may be that it does not make much sense to establish time relationships from phonocardiograms with an accuracy of more than 5 milliseconds.

List of Symbols

λ,μ	=	Lamé elastic constants
G	=	shear modulus
ν	=	Poisson constant
ω	=	circular frequency
k	=	wave number in compressional wave propagation
h	=	wave number in shear wave propagation
c_l	=	propagation velocity of compressional waves
c_l	=	propagation velocity of shear waves
ρ	=	density
α_l,α_t	=	attenuation constants in compressional, respective shear waves
r	=	distance from a point to the origin
e	=	base of natural logarithms
i	=	$\sqrt{-1}$

Glossary of Terms

Mechano-electric transducer: Device which converts a mechanical variable into an electric one

Visco-elastic medium: A medium having both elastic and viscous properties

Dispersion: Frequency dependence of propagation velocity

Attenuation: Diminishing of amplitude of waves

Accelerogram: Recording of acceleration

Compressional wave: Wave caused by compression of the medium; in a plane wave displacements are parallel to the propagation direction (longitudinal waves)

Shear wave: Wave caused by shear of the medium; in a plane wave displacements are perpendicular on the propagation direction (transverse waves)

Mechanical impedance: An expression in vibration mechanics analogue to electric impedance defined by Z_M = Force/velocity

Piezo-resistive material: Electric resistance of the material changes if a stress is applied to it

Piezo-electric material: Gives an electric tension if a stress is applied to it

References

Abramowitz M, Stegun I A: Handbook of mathematical functions. Dover Publications, New York 1968

Chandraratna P A N et al: Echocardiographic observations on the mechanism of production of the second heart sound. Circulation 51, 292 (1975)

Craige E: On the genesis of heart sounds, contributions made by echocardiographic studies. Circulation 53, 207 (1976)

Cheung Y S: A feasibility study of the spatiotemporal analysis of cardiac precordial vibrations. Thesis, University of London 1977

Dunn F, Fry W J: Ultrasonic absorption and reflection by lung tissue. Phys Med Biol 5, 401 (1961)

Einthoven W: Die Registrierung der menschlichen Herztöne mittels des Saiten-Galvanometers. Pflügers Arch Ges Physiol 117, 461 (1907)

Faber J J, Burton A C: Spread of heart sounds over the chest wall. Circ Res 11, 96 (1962)

Faber J J: Damping of sound on the chest wall. Circ Res 13, 352 (1963)

Feruglio G A: An intracardiac sound generator for the study of the transmission of heartmurmurs in man. Am Heart J 63, 232 (1962)

Franke E K: The propagation of surface waves over the human body. U.S.A.F. Tech Rpt, No 6494, 1951

Fung Y C: Stress-strain history relation of soft tissues in simple elongation. In: Biomechanics its foundations and objectives, chapt 7. Prentice Hall, New York 1972, p 192

Gierke H E von et al: Physics of vibration in living tissues. J Appl Physiol 4, 886 (1952)

Groom D: Standardization of microphones for phonocardiography. Biomed Eng 5, 396 (1970)

Heintzen P, Viets K W: The diacardiac phonogram. Am Heart J 65, 59 (1963)

Hök B et al: Imaging of chest motion due to heart action by means of holographic interferometry. Med Biol Eng 16, 363 (1978)

Kozmann G: An acoustical model of the heart. In: Medical Computing, Proceedings Int Symp IRIA (M Laudet et al, eds). Taylor and Francis, London 1977, p 253

Lambert R K, Wilson T A: A model for the elastic properties of the lung and their effect on expiratory flow. J Appl Physiol 34, 43 (1973)

Landau L D, Lifshitz E M: Theory of elasticity. Pergamon, Oxford 1970

Lewis D H: Phonocardiography. In: Handbook of physiology, Sect 2. Circulation. Vol. I, chapter 22, p 695. Am Physiol Soc, Washington 1962

Luisada A A: From ausculation to phonocardiography. Mosby Company, St. Louis 1965

Luisada A A et al: Changing views on the mechanisms of the first and second heart sounds. Am Heart J 88, 503 (1974)

MacCanon D M et al: Mass excursion parameters of first heart sound energy. J Appl Physiol 27, 649 (1969)

Mathews F L, West J B: Finite element displacement analysis of a lung. J Biomech 5, 591 (1972)

Mc. Kusick V A: Cardiovascular sounds in health and disease. Williams and Wilkins, Baltimore 1958

Morse P M: Vibration and sound. McGraw Hill, New York 1948, p 316

Netter F M: Ciba collection of medical illustrations. Circulatory System, Vol. 5, 1969

Oestreicher, H L: Field and impedance of an oscillating sphere in a visco-elastic medium with an application to biophysics. J Acoust Soc Am 23, 707 (1951)

Ravin A, Frame F K: International bibliography of cardiovascular ausculation and phonocardiography. Am Heart Ass, New York 1971

Remington J W: Tissue elasticity. Washington 1957

Rodbard S, Mathews N P: Valve closure and second heart sound. Circulation 52, 519 (1975)

Rushmer R F: Cardiovascular dynamics. Saunders, Philadelphia 1961

Sakomoto T et al: Hemodynamic determinants of the amplitude of the first heart sound. Circ Res 16, 45 (1965)

Scher A: Mechanical events of the cardiac cycle. In: Physiology and Biophysics. (Eds. T. C. Ruch and H. D. Patton) Chapter 19. Saunders, New York 1966, p 550

Shah P M et al: Spread of the mitral sound over the chest. Circulation 28, 1102 (1963)

Verburg J, Strackee J: Phaseless recursive filtering applied to chest wall displacements and velocities using accelerometers. Med Biol Eng 4, 483 (1974)

Verburg J: Fysische aspekten van de transmissie van harttrillingen. Thesis, Amsterdam 1975

Verburg J, Vollenhoven E van: Phonocardiography, physical and technical aspect and clinical uses. In: Non Invasive Physiological Measurements (ed. P. Rolfe). Acad Press London 1979

Waider W, Craige E: The first heart sound and ejection sounds: echo-phonocardiographic correlation with valvular events. Am J Cardiol 35, 346 (1975)

Wells P N T: Physical principles of ultrasonic diagnosis. Acad Press, London 1969

Zalter R et al: Acoustic transmission characteristics of the thorax. J Appl Physiol 18, 428 (1963)

J. Verburg,
Laboratory of Medical Physics,
Herengracht 196,
NL-1016 BS Amsterdam
(The Netherlands)

Adv. cardiovasc. Phys., vol. 5 (Part III), pp. 104–116 (Karger, Basel 1983)

Structure and Performance of a Measurement Program for Systolic Time Intervals and Phonocardiography

C. L. Tuinstra
State University Hospital, Leiden, The Netherlands

Contents

I. Abstract

A computer program is described which can determine the presence of different phenomena in the phonocardiogram, which measures the systolic time intervals and which indicates abnormal observations.

The program compares measurements from the ECG (lead II), the right carotid artery pulsations and the phonocardiogram taken at the second intercostal space to the right (2R), and third intercostal space to the left of the sternum (3L) and at the apex. It determines the pre-ejection period (PEP), left ventricular ejection time (LVET) and the distance from the onset of the QRS complex to the first and to the second sound, the presence of systolic and diastolic extra sounds (ejection click or opening snap), the presence of protosystolic, protodiastolic and late diastolic murmurs, as well as holosystolic, holodiastolic and continuous murmurs. It measures the splitting of first and second heart sounds as well.

This program is meant to be used in hospitals as an aid to auscultation and phonocardiography on a routine base and it is an attempt to quantify and specify certain traumatic findings.

II. Introduction

When, in about 1960, the first analog to digital converters became available for analysis of physiological signals, investigators in many laboratories started to develop programs to automate the measurements and the classification of electrocardiograms, electroencephalograms and phonocardiograms (PCG).

The development of ECG-analysis programs has been rather successful and indeed a number of programs are available to the cardiologist and the epidemiologist for application in the clinic or for screening purposes.

Although the ECG is a rather well defined signal, that originates from known processes, one is surprised at the number of controversies and disappointments which arose during the infancy of this application area (Zywietz and Schneider 1973). There still are many different systems in use for ECG wave detection and ECG-classification and the developmental work is still in progress.

Working on phonocardiograms one may expect to encounter even more problems than those encountered in the electrocardiograms. Lack of calibration and lack of quantitative descriptions in general indeed have obstructed the progress in the development of these phono-programs. To date, there exists only one program that is of more than local importance (Yoshimura et al 1973).

Many authors have reported on programs to measure systolic time intervals (Kyle et al 1970; Swatzell et al 1973) however, progress on the interpretation of the sounds has been rather limited. This could be due to the idiosyncrasies of the PCG, to the lack of standardization of equipment, to the wide inter observer variation in PCG reading or also to the vulnerability of the PCG to artifacts. But if technically representative tracings become obtainable, the PCG can provide us with important information on the functioning of valves and great vessels and in an indirect way on the heartmuscle. A program for automated analysis of the PCG might also be of importance as an aid to the existing auscultatory techniques and, particularly in combination with automated ECG analysis, it might be useful in screening populations for heart disease, where non-invasive techniques are mandatory.

III. Program Philosophy

In the development of the program one has to keep in mind the following requirements:

- the program should be independent of the computer type, which means that it should be written mainly in a high level language, e.g. FORTRAN.
- the program should be independent of the recording equipment and independent of the recording technicians.
- the program should need no manual interference after the recording and hence it should not be interactive.
- the program should work independently of any calibration of the amplitudes in the signal, because there are no reliable methods available to calibrate the registration.
- the program should be easy to modify or to expand and hence it should be modular in design.

In the design of a PCG program it is very tempting to tackle the stochastic properties of the PCG with mathematical methods which are known from other fields of pattern recognition like Fourier analysis, cepstrumcalculations, auto- and crosscorrelation functions. Indeed these methods can be applied to the PCG, but one will not find it very rewarding, because one cannot satisfactorily classify the normal and pathological PCG's in the terms of any of these methods.

For this reason it was impossible to abstract new clinical information from the PCG with these methods that could not be seen by a trained cardiologist by conventional reading.

The program described here uses an empirical approach, it describes the PCG in terms of sounds and murmurs and relates the occurrence of certain phenomena to timing points found in the ECG and the carotid-pulse tracing. The development is hampered by the lack of definitions which are adequately precise to be employed in the detection algorithms of the sounds and murmurs.

IV. Methods

Simultaneous tracings of the ECG (lead II), right carotid artery pulsation, PCG (with Maas Weber filter 140 and 75 Hz) were recorded on analog magnetic tape in three runs, with the microphone successively in the second

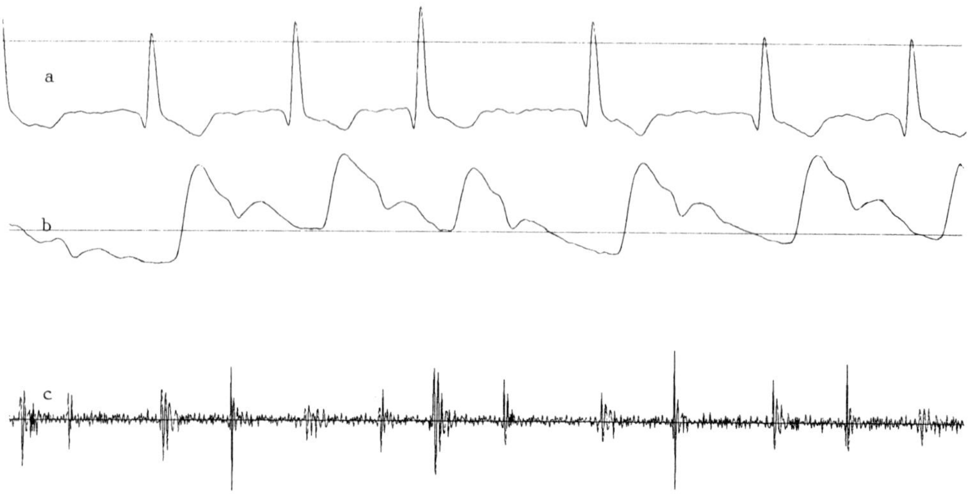

Fig. 1 a ECG lead II; b Carotis dextra pulsation; c Phonocardiogram.

intercostal space right of the sternum (2R), in the third intercostal space left of the sternum (3L) and at the apex, found by palpation with the patient in the supine position. The recordings were taken by the technicians responsible for the PCG tracings routinely used in the clinic. The average quality therefore was as good as can be expected in a clinical setting. The tracings are shown in Fig. 1.

The ECG and carotid pulsation were digitized with sample-frequency 300 Hz, the PCG with 1000 Hz per channel, six seconds in each run.

- ECG

  The onset of the QRS complexes was determined by an algorithm based on a set of level criteria depending on the signal to noise ratio in the first derivative.

  The exact algorithm is described in appendix A.

  Carotid artery pulsation

  The foot of the upstroke of the carotid pulsation and the incisurae are found using an algorithm based on a second derivative of the smoothed signal. The shape of the incisura is expressed as the difference of the slopes before and after incisura.

  This algorithm is provided in appendix B. The amplitudes of the carotid pulsation in the different heartcycles in one tracing are compared with a threshold, any pulsation smaller than that threshold is rejected from the measurements, being a suspected premature contraction. The threshold is set as 30 % of the maximum amplitude in that tracing.

  PCG

  In the measurements of the PCG we used the envelope of the rectified signal.

  The PCG does not provide any markings of its own, like QRS-waves of the ECG do, or are provided by the steep upstroke in the carotid artery which can determine the phase of the heartcycle. Neither do we have any amplitude calibration available. All measurements therefore have to be done relative to some standard, found in the signal itself.

  The amplitude thresholds used in search for sounds and murmurs are related to the first or second sound and the searching intervals are dependent on the R-R distance.

  An algorithm based on a set of time-amplitude criteria determines the onset and termination of the first and second sound, where the amplitude criteria are related to the amplitude of the first sound. The algorithm aims at finding a silent zone between the maximum of the first sound and the maximum of the second sound. If no silent zone is found, the program concludes a holosystolic murmur and the threshold is raised. In searching for the onset and termination of the sounds, holosystolic, holodiastolic and continuous murmurs are found;

using the same "silent zone" algorithm, systolic or diastolic extra sounds (ejection click, opening snap) are located.

More details about this silent zone algorithm and its application to the first and second sound are given in appendix C.

If the beginning and end of the sound are located but give improbably broad sounds, adjacent murmurs are supposed to be present. In combination with the location of the maximum of the sound we can distinguish proto- and late systolic murmurs and proto- and late diastolic murmurs as described in appendix C.

An important phenomenon in heartsounds is often referred to as "splitting". It occurs when a heartsound shows two clearly descernable maxima and it is often seen in the second heartsound. This sound consists of two parts: the aortic part (closing of the aortic valves) and the pulmonary part (closing of the pulmonary valves). These parts often coincide, but normally the pulmonary part is seen slightly (30 ms) after the aortic part, this distance is depending on the respiration. Splitting is abnormal if the distance is more than 70 ms or if the pulmonary part *precedes* the aortic part.

Both first and second sounds are investigated for splitting; abnormally wide splitting or "reversed" splitting of the second heartsound can be detected by the program. (Fig. 2)

To *suppress artifacts* as much as possible all found phenomena are compared in five heartcycles; a majority rule decides if a noise or sound is to be called an artifact or not.

The output is confined to a report of the findings in the signal, there is no interpretation of these findings because we have no clearcut decision scheme, as yet.

But a number of measurements, along with the indication of some possible abnormal symptoms can be of help to the practitioner who must himself decide about the interpretation of these symptoms.

V. Performance

The reliability of the measurement program was tested by comparing the location of the points found by the program with those determined by a cardiologist (H. Koster). The points of interest to us are the onset of the QRS complex in the ECG, the onset of the rapid upstroke of the carotid artery pulsation and the beginning and termination of the first and second heartsounds.

The records were replayed by the computer with a paperspeed equivalent to 50 mm/s. From paperspeed, line-thickness and repeatability of the cardiologist we calculated an uncertainty of about 20 ms inherent to the testing procedure. The repeatability of the cardiologist was in the determination of the onset of the QRS slightly better (15 ms) than in the determination of the onset of the upstroke of the carotid artery pulsation or the beginning and end of the heartsounds (20 ms). The results of these tests are shown in Figs. 3–8.

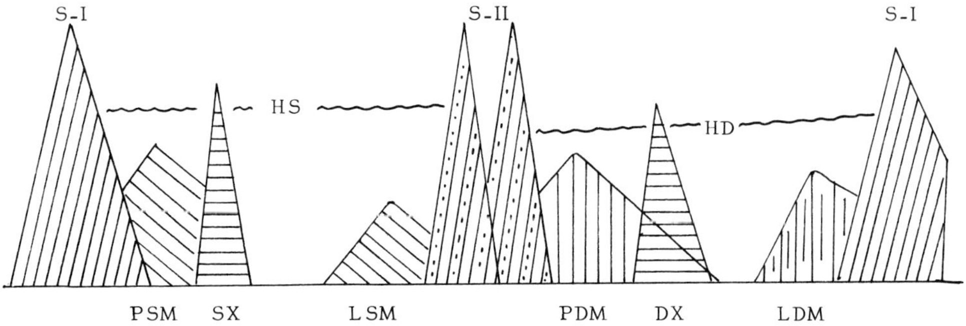

Fig. 2 Phenomena which can be found in each heartcycle. S-I = first heart sound; S-II = second heart sound; PSM = protosystolic murmur; SX = systolic extra sound; LSM = late systolic murmur; HS = holosystolic murmur; PDM = protodiastolic murmur; DX = diastolic extra sound; LDM = late diastolic murmur; HD = holodiastolic murmur.

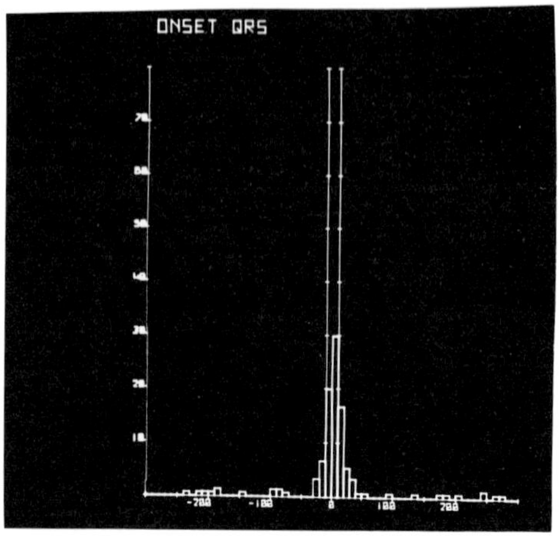

Fig. 3

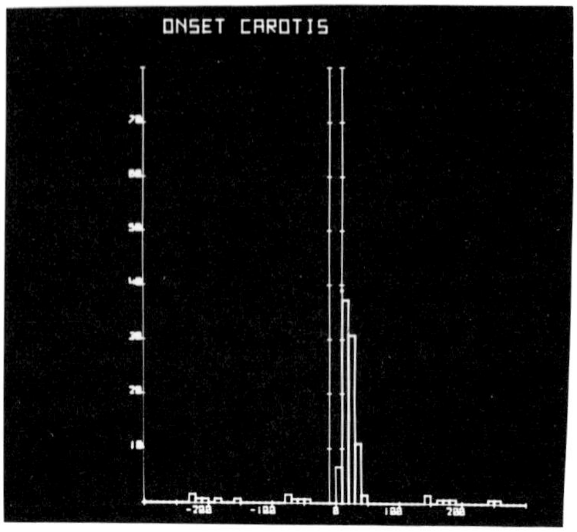

Fig. 4

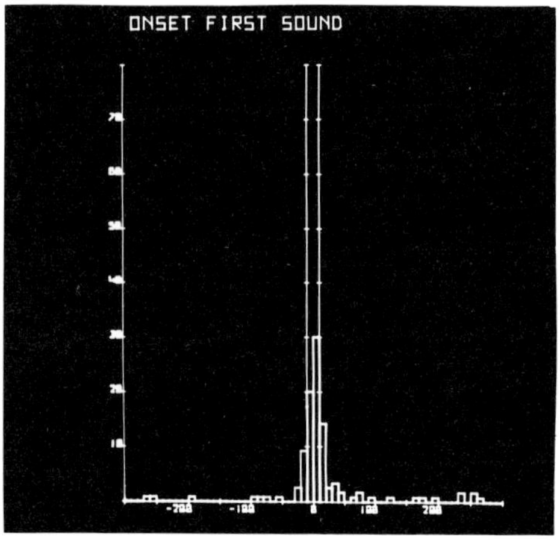

Fig. 5

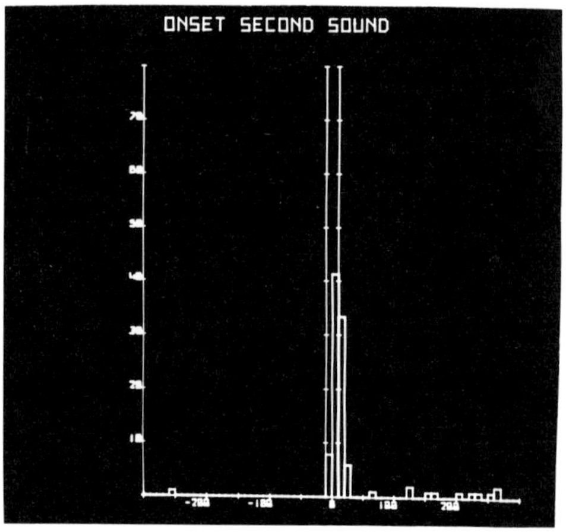

Fig. 6

In these bar-graphs measurement-differences between program and cardiologist are shown. In the centre, the difference is smaller than the precision of the comparison-procedure. This area is indicated by the two vertical gridlines in the centre. All measurement differences which lie between those gridlines can be assumed to be negligible, at least the program measurements agree with the "manual" measurements.

From Fig. 3 we can conclude that 76% of the locations of the Q-onset are found within 20 ms. We must regard (from this magnitude of error) the program as too inaccurate, because of the fact that the location of the onset of Q is used in the calculation of the pre-ejection period (defined as the Q-S2 distance minus the left ventricular ejection time). The PEP is of the order of magnitude of 100 ms, so errors of 20 ms are intolerable. In general, the determination of the Q-onset in a single ECG lead is questionable, because it is not sure that we can see the earliest part of the depolarization in that particular lead. The use of three leads is therefore preferable.

From Fig. 4 we can conclude that the program finds the foot points in the carotid systematically a little early and that the average of this shift is 20 ms. If we compensate for this shift by adding 20 ms to the points which were originally determined as the onset of the rapid upstroke, 87% of all points are found within 20 ms, when compared to the points manually indicated by H. Koster.

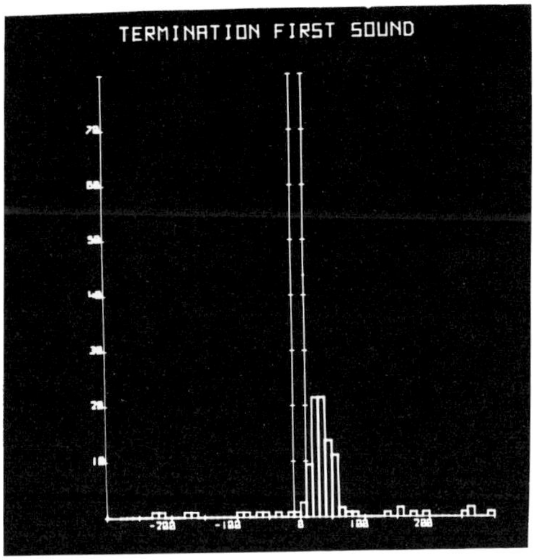

Fig. 7

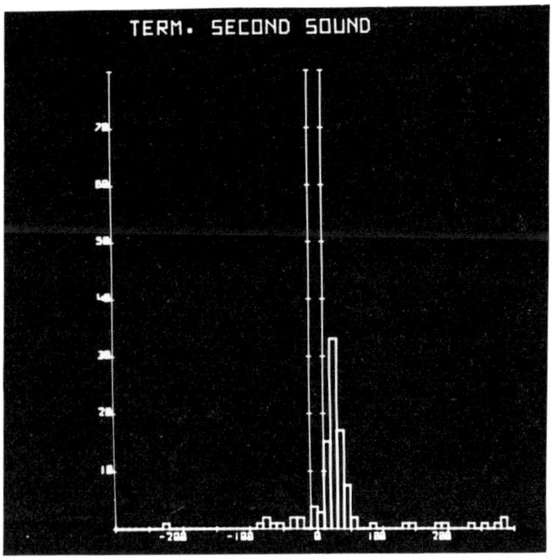

Fig. 8

The manual indication of the beginning and termination of the heartsounds is difficult because all kinds of vibrations are blurring the borders of the heartsounds. In this light the accuracy of the location of the beginning (Figs. 5 and 6) and termination (Figs. 7 and 8) of the heartsounds is surprisingly good. The onset of the first sound is found in 76% of the cases, of the second sound in 86%, the termination of the first sound in 68% and of the second sound in 75% of the cases within ± 20 ms of the manually determined points.

In the termination of the heartsounds (Figs. 7 and 8), a systematic error of 30 ms is found in both sounds. Apart from this, the results of this test do not give reason to change the algorithm.

Tracings with moderate artifacts were not excluded from the tests, because we wanted to know the performance of the program under real live circumstances. About 80% of all errors greater than 40 ms could be traced to be due to artifacts.

As regards extra sounds and murmurs it is impossible to present hard figures about the reliability of the program since the existence of these phenomena depends on questions like: what exactly do we call an extra sound? How are murmurs defined numerically? As judged by the clinical readers our approach gave quite satisfactory results, though it should be stated that it cannot compete with a trained cardiologist.

Other errors are that small murmurs are missed and that sometimes the opening snap is fused with the second sound and therefore is not detected. In general problems arise if the maximum amplitudes of the first and second sounds are extremely different, for instance if the amplitude of the second heartsound is three times greater than the amplitude of the first sound. In that case the algorithm to find onset and termination of the sounds (described in appendix C) starts with a threshold which is too high and it "overlooks" the smaller murmurs.

In principle, this could be improved by making the initial threshold a function of the amplitude ratio of first and second sound. And because it is known from auscultation that the first sound is much less audible at the apex than at 3L or 2R, the threshold could also be a function of the location of the recording on the thorax.

But in our recordings we did not find any significant difference in the amplitude ratio's, therefore we did not implement this in the program as yet.

The effect of noise in the signal was determined by comparing the results of a random sample of tracings with a subset of selected, good quality tracings. It was concluded that the differences are small indeed. The fact that the program performs well in tracings of mediocre quality, in fact in the same way as it does in tracings of good quality, is rather satisfactory.

VI. Application

The program was applied to a set of pulse tracings and phonocardiograms of 130 volunteers, who had no history of heart-disease to measure the normal ranges of the time intervals as measured by this program.

VII. Appendix A

Algorithm to locate the onset of the QRS complex (Fig. A1).

The ECG data samples should be stored in an array in the main memory. It is tested for sample frequencies between 200 and 500 Hz. The total observation time as contained in the sample array in the memory should be at least long enough to contain two heartbeats.

The QRS complexes are localized by finding the relative maxima of the first derivative greater than a certain threshold. This threshold is 60% of the absolute maximum of the first derivative. A minimum distance between the complexes of 300 ms. is assumed, this implies that at heartrates over 200 beats per minute some QRS complexes are not found correctly.

Over the first 100 msec of each QRS complex the average absolute derivative is computed as a noise threshold. The onset of the Q is defined as the first maximum not trespassing this threshold after the first minimum under this threshold taken from the maximum of the absolute derivative.

VIII. Appendix B

Algorithm to find onset of the rapid upstroke and insicura of the carotid artery pulsation (Fig. B1).

First find maximum derivative (positive slope) in the signal. Try to find several other points where the first derivative is close to this maximum. However, if these points are followed within 30 ms by a steep negative slope, these are probably artifacts. Once the high frequency noise is eliminated the absolute maxima in the positive slope are located. The preceding minima (base) and following maxima (tops) are determined. The onset of the rapid upstroke is defined at the point from where, searching from base to top, two consecutive samples are greater than the mean slope between base and top.

The insicura is defined as the point between top and following base where the maximal difference between the sample points and a line from the top of the pulsation to halfway the following Q-wave.

IX. Appendix C

Application of the silent zone algorithm to the localization of onset and termination of heartsounds.

The silent zone algorithm (SZA) is a general algorithm, which can be used to determine the end of a burst of extra noise in a noisy signal (Fig. C1). It uses input parameters (A) and (B) to define begin and end of the observation period in which it should find the end of the burst; it uses input parameters (T) and (Z) to define an amplitude threshold and the length of the silent zone (Z A-B). A zone is called silent if no noise peaks are present greater than T. The result is given in output parameter (E), if no silent zone is found E is set to zero.

The SZA is applied to the PCG to locate the heartsounds. The parameters A, B and Z are dependent on the R-R intervals. The threshold T has a starting value, which depends on the maximum amplitude in the PCG during that heartcycle. Fig. C2 is a flowchart which illustrates the routine to locate the first heartsound. It uses the location of the QRS complex input parameter (R(N)) and output parameters BFSND, MFSND, TFSND and CONTS, which indicate the begin, maximum and termination of the first sound and the intensity of a continuous murmur. Fig. C3 illustrates the localization of begin, maximum and termination of the second sound. This routine uses input parameters (R(N)) and the location of the incisura of the carotid pulsation (I(N)); the output indicates also the intensity of the systolic and diastolic murmurs.

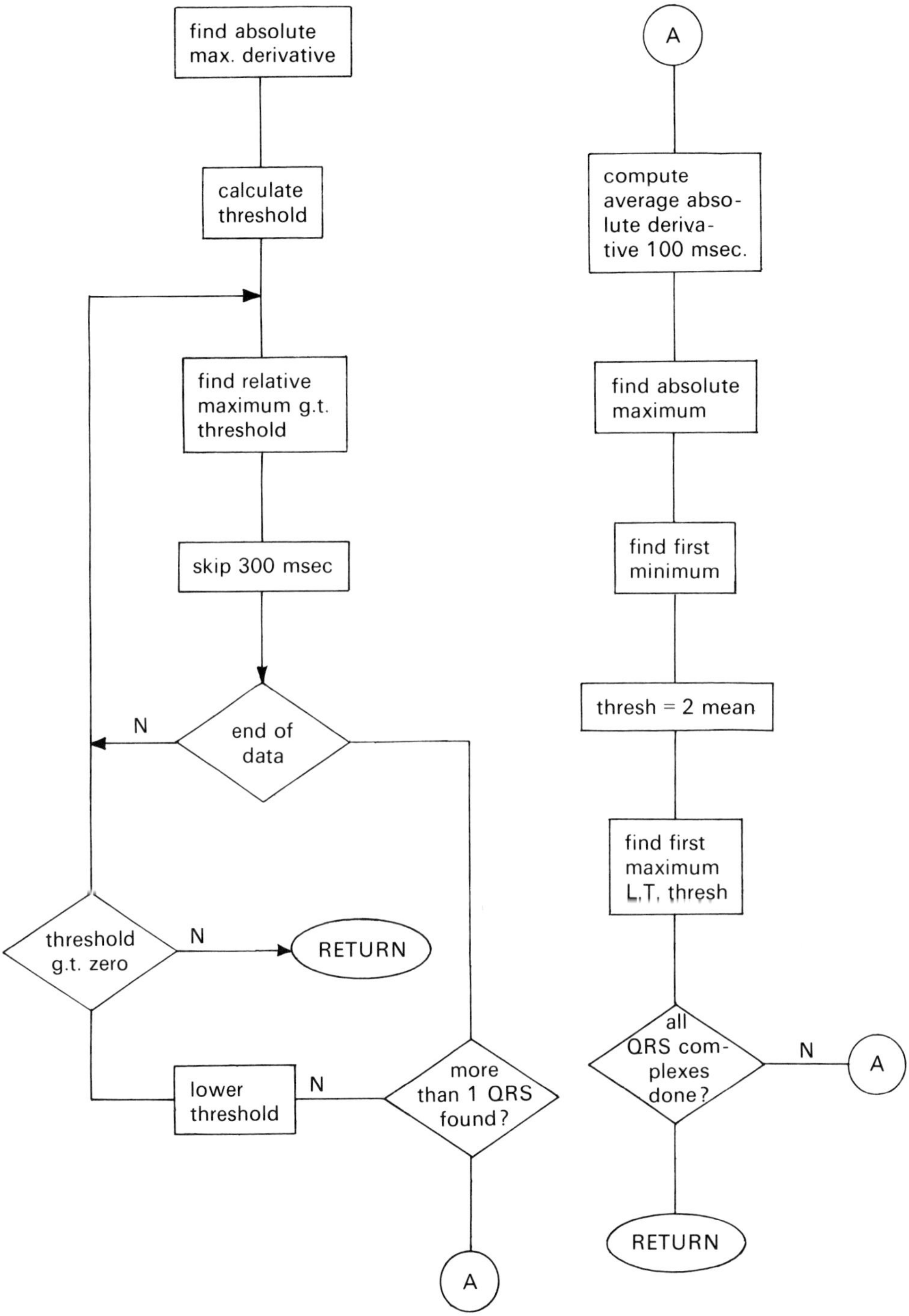

Fig. A1. Flowchart of the algorithm to locate the onset of the QRS complex.

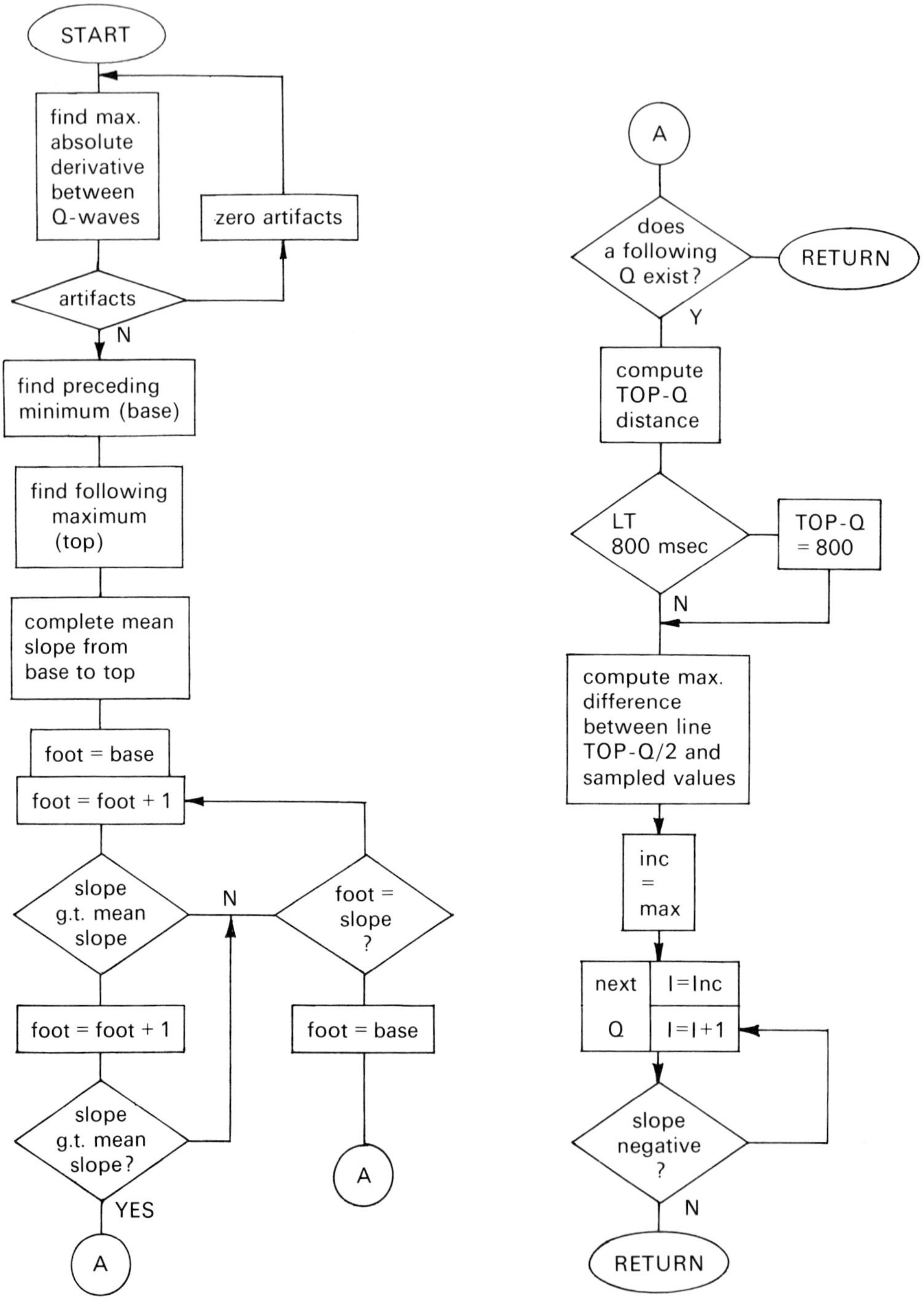

Fig. B1. Flowchart Carotid Artery Pulsation.

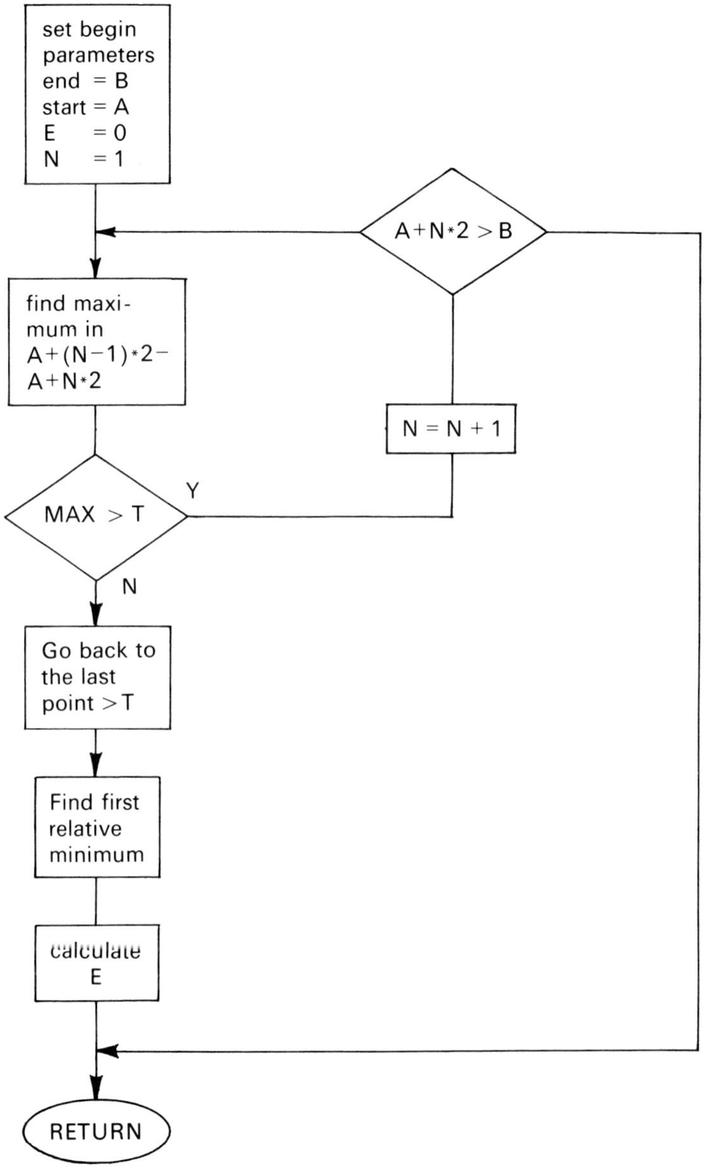

Fig. C1. Flowchart silent zone algorithm.

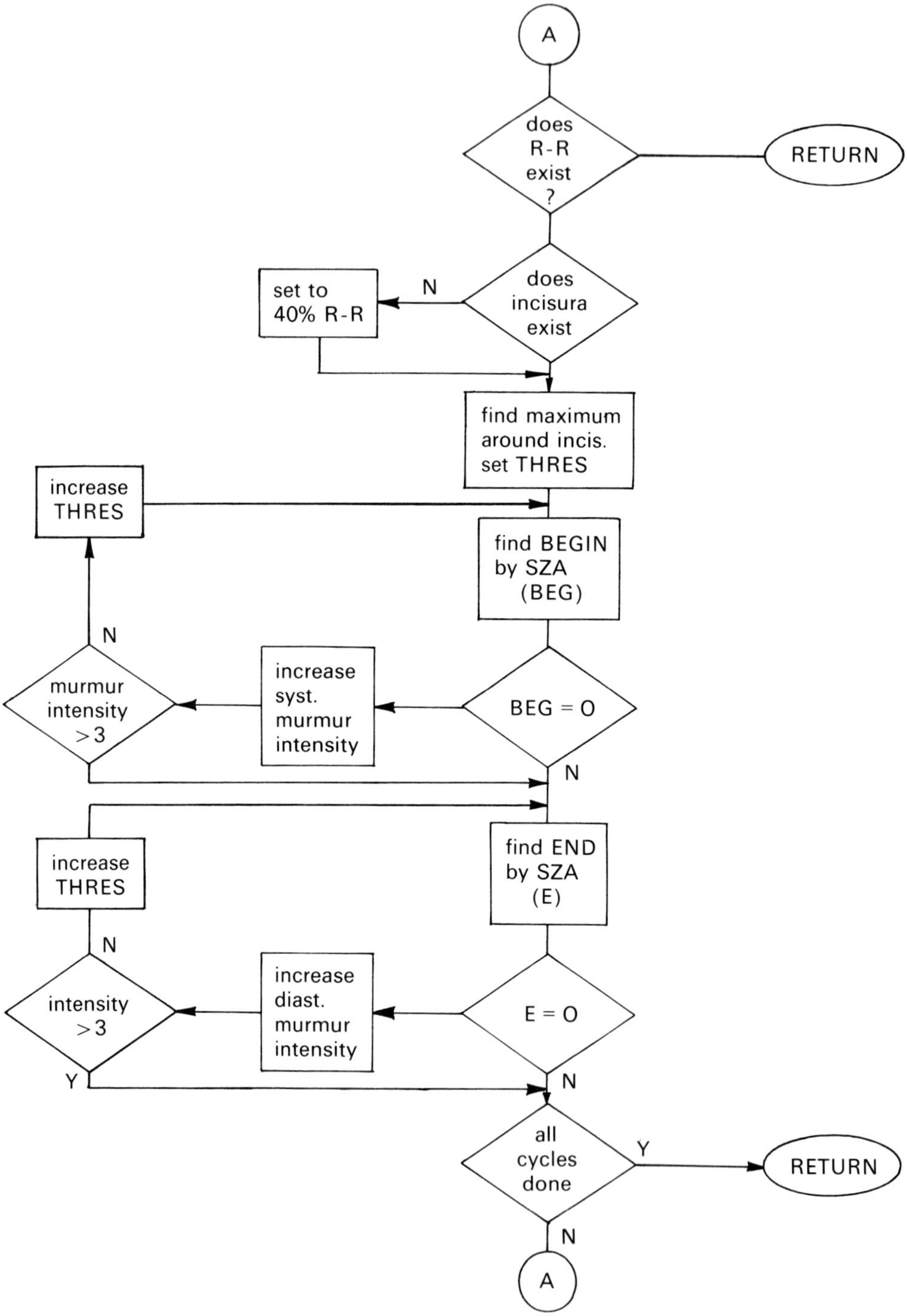

Fig. C2. Flowchart of the program to determine begin and end of the first heartsound.

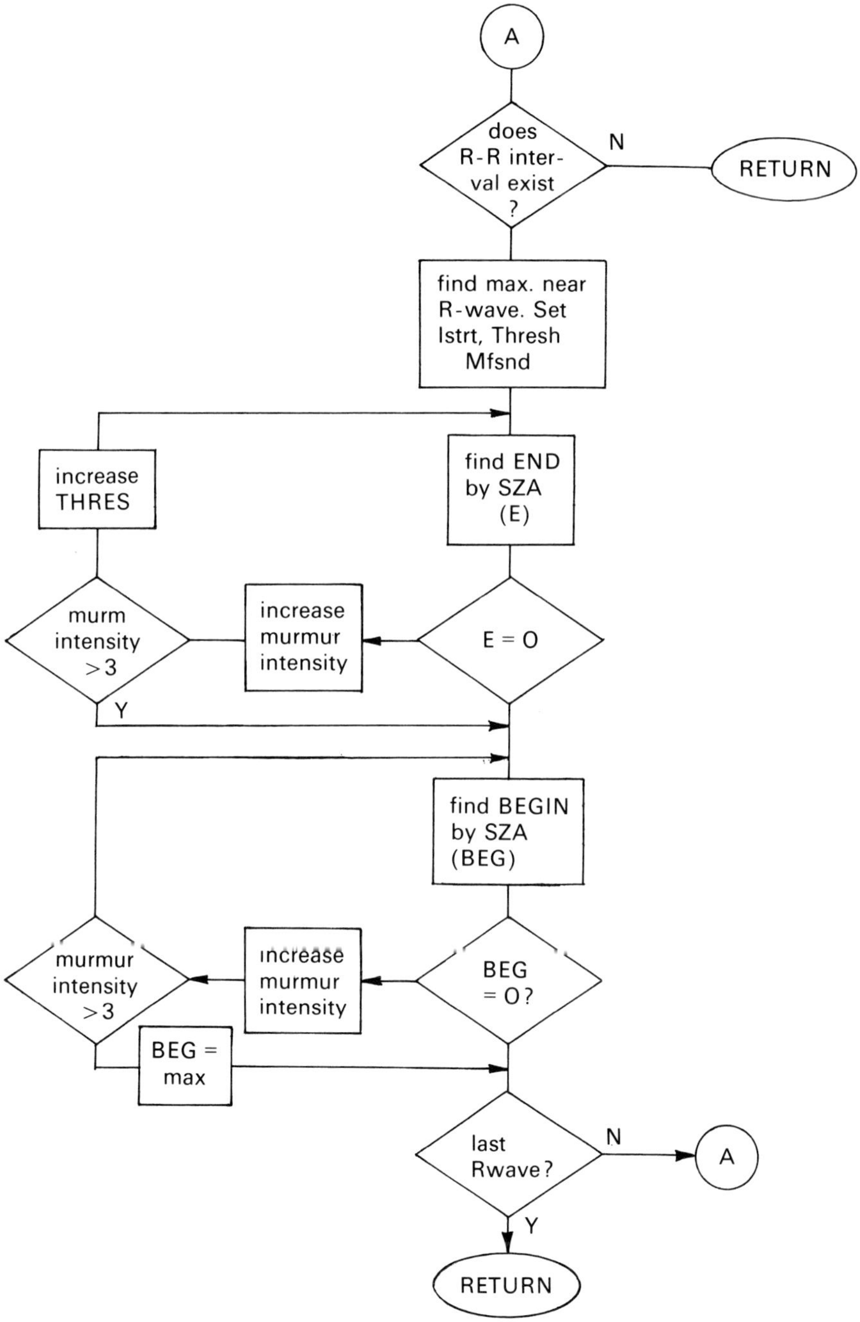

Fig. C3. Flowchart of the program to determine begin and end of the second heartsound.

References

1. Kyle M C, Freis E C: Computer identification of systolic time intervals. Comput Biomed Res *3*, 637 (1970)
2. Swatzell R H, Bancroft W H Jr, Macy J Jr, Eddleman E E Jr: The on-line computer system for determinining the systolic time intervals. Comput Biomed Res *6* (5), 465 (1973)
3. Yoshimura S, Shimoj E, Nakatsuka T, Ohara M: Automatic screening system for the cardiac diseases of school children. Digest of the 10th International Conference on Medical an Biological Engeneering, Dresden (1973)
4. Zywietz C, Schneider B (eds): Computer application on ECG and UCG analysis. Proceedings on the 2nd IFIP TC-4 Workshop on Computer Application on ECG and VCG Analysis (1973)

Dr. C. L. Tuinstra,
Dept. of Medical Information,
State University Hospital,
Boerhaavekwartier,
Leiden
(The Netherlands)

Adv. cardiovasc. Phys., vol. 5 (Part III), pp. 117–156 (Karger, Basel 1983)

Evaluation of Medical Waveforms by Computers

I. Tomek
ACADIA University, Wolfville, Nova Scotia, Canada

Contents

I. Introduction

Automation of processing of medical waveforms such as the electrocardiogram has a relatively long history. This is natural since computers are perfect replacements of humans wherever manipulation of numerical and logical values has to be performed and this seems to be a good description of what a human evaluator does when reading the ECG. Highly qualified medical personnel, performing these routine tasks, could be freed to devote more of their time to patients if computers could replace them in time consuming evaluation of records.

The problem of automatic diagnosis was initially attacked by trying to imitate the routine followed by cardiologists. Programs developed by this approach achieved a large degree of perfection. Their performance is comparable to that of humans. Some of them are used in clinical practice. These imitative approaches have gradually been complemented by more sophisticated methods based on results from mathematical statistics (more specifically its subfield – statistical pattern recognition). These approaches have not yet attained the degree of acceptance they deserve, for reasons to be discussed later. To change this situation is one of the goals of people working in this field. At the same time, attempts to attack a deeper problem associated with waveform evaluation, that of the understanding and at least partial algorithmization of the process by which humans discover criteria which they use for evaluation have also been initiated. This exciting subject with enormous practical importance is, however, much more complicated and it will probably take a long time before practically useful solutions are found.

In this chapter, we will describe waveform processing in stages, from the point of view of statistical pattern recognition. We will show the sequence in which programs, for waveform classification, are (or could be) prepared and discuss the individual steps in some detail. We will then show on examples of specific applications some results of the research undertaken during the last decade.

The approach taken in this chapter is rather different from that used in most of the voluminous literature on the subject of computerized evaluation of medical waveforms*. This subject is usually studied from the point of view of a specific application (for example the ECG) and the approach very often is to translate into an algorithm the routine used by a trained specialist in a similar situation. We are going to use the abstract point of view of a person faced with the task of designing an efficient algorithm for automatic evaluation of records of a certain type. In fact we do not even assume that we are dealing with medical waveforms. This approach is used for several reasons: This chapter as the rest of the book, attempts to define and describe a problem in quantitative and algorithmizable terms. Herein we are trying to describe generally applicable methods rather than specific partial solutions. Although much work has been done on automatic evaluation of medical waveforms, the theoretical knowledge accumulated in statistical pattern recognition has not been fully used (most programs use heuristic approaches rather than methods with mathematically proven efficiency). This is partly due to the fact that until recently much of the modern research in pattern recognition could only be abstracted from specialized journals and understood by specialists, since no up-to-date textbook on this subject existed. In the last few years this situation has changed drastically with the publication of a number of excellent books on pattern recognition. None of them, however, contains material devoted specifically to waveform processing. It is hoped that this chapter will provide the interested reader with a systematic outline of the subject, stimulate his interest and induce him to read some of the more detailed surveys pertinent to the subject and to apply them to the analysis of physiological waveform data.

This chapter does not deal at all with the related problem of picture processing such as encountered for example in the processing of X-ray pictures. Some problems and their solutions are different although the essential stages of the process are identical. Textbooks on pattern recognition (for example Duda and Hart (1973)) touch on this topic and are recommended as an introduction for the interested reader. However, a detailed discussion of this field is beyond the scope of this chapter.

Another important problem not treated in detail in this chapter is the conversion of signals from the analog to the digital form. This is because A/D conversion is largely a hardware

* An exception is the book by Young and Calvert (1974) on pattern recognition, which contains a chapter on ECG processing, written from the point of methodology of computerized diagnosis rather than the specific problem of ECG processing.

problem, which is not the subject of this chapter. As in the whole process of design of an algorithm for signal evaluation, there is a considerable amount of feedback between the choice of an A/D converter and the following stages. Indeed certain basic knowledge of the signal to be processed is required for a satisfactory design of an A/D converter, namely the frequency range of the useful part of the signal which is essential for the choice of sampling rate, and the amplitude range which together with information about the noise present in the signal determines the number of levels into which the signal will be quantized. (A short outline of the principles of A/D conversion is given in Appendix 2.)

Also we are not going to discuss other circuits which serve to modify the original signal, such as amplifiers and electronic filters. These circuits are very important parts of signal processing. They are used mainly to amplify too weak signals and eliminate undesirable frequency components (noise etc.). Occasionally they are also used to integrate or differentiate the original signal. So, for example, displacement ballistocardiograms may be electronically differentiated to obtain acceleration ballistocardiograms and vice versa. For a detailed treatment of the function and design of amplifiers and filters the reader may consult for example the book by Smith (1971).

Although the presentation of this chapter is dictated by the goal of waveform classification, certain parts have their own independent importance. This is particularly true for the section dealing with preprocessing: Digital filtering, for example, can be used to smooth out a signal even when it is to be evaluated only by a human operator. Data reduction by piecewise-linear approximation is advantageous for the reduction of the required volume of storage even when no further computer processing is contemplated. The same is true for extraction of typical complexes.

II. Systematic Design of Diagnostic Programs Using Methods of Pattern Recognition

Pattern recognition in its classical form is a discipline derived from mathematical statistics. It studies classification of random samples and attempts to give algorithmizable solutions amenable to computer programs. The range of applications is very broad: the recognition of printed characters, recognition of fingerprints, evaluation of ECGs, automatic processing of aerial photographs are a few examples. We will show what pattern recognition does in medicare delivery:

Let us assume that we have a large collection of classified electrocardiograms, recorded in digital form. Each is labeled as being an ECG of either a healthy subject or that of a subject with an abnormal heart. We want to design a fully automatic process capable of classifying any new ECG into one of the two classes: normal or abnormal or, possibly, an undecidable borderline case. That part of this process which is concerned with designing an algorithm for efficient classification based upon the available data is in the domain of pattern recognition. It is obvious that the nature of the problem is statistical: individual normal records (samples) are not identical – they vary with the age of the subject, the condition of his heart (the limits between normal and abnormal are vague) etc. There is a degree of randomness associated with the shape and amplitudes of individual records. The two classes (normal and abnormal ECGs) in fact overlap when they are generated on the basis of information other than the ECG: a number of healthy subjects may have an ECG very close to that of a subject with abnormal circulation and vice versa. In short each record is a sample from one of two generally overlapping probability distributions. (For a qualitative outline of the basic statistical terminology necessary for the understanding of the rest of this chapter, the reader is referred to Appendix 1.)

Before starting an outline of the methods used in the process of designing a diagnostic program, let us stress that the very limited size of this chapter makes it impossible to discuss even certain fundamental methods. So, for example, discriminant analysis is mentioned in the last section on diagnostic programs but never explained in the main text. Information on these subjects can be found in Meisel (1972) and other books dealing with statistical pattern recognition.

A. The Stages Involved in the Design of a Classifier

We will now describe the stages involved in the design of an efficient algorithm for waveform classification:

1) Selection of waveform features.

At the beginning, we have to examine the set of all those classified records that we want to use for the design of our algorithm (the "design set") and search for suitable features which can be used as indicators of class membership. Such features may, in the case of ECG classification, be values of variables derived from the P, Q, R, S and T waves such as amplitudes, durations and slopes. Let us mention that it is desirable to keep a part of our classified data for the purpose of evaluation of the efficiency of our program, this part of the set of classified samples is the "test set" and is not usually used as a part of the design set.

2) The program: signal preprocessing, feature recognition and classification.

When we think that the chosen set of features is sufficiently large and has enough discriminatory power, we can start working on our program. The program will in general consist of the following parts:

1. Preprocessing of the Input Signal and Recognition of Features

The original record as obtained from the measuring device is noisy. Some of the noise (that component which is in a frequency range separated from the range of the useful signal) can be filtered out. This may be, for example, the 60 cps noise originating from electrical components, vibrations from the floor (encountered for example in ballistocardiograms), muscle tremor etc. Filtering will be performed if the useful signal does not have important components in the same frequency range as the noise: Linear filtering attenuates and distorts components of a signal which belong to a chosen range of frequencies. There is no way of removing noise without distorting the useful signal if both have the same frequency. Elimination of particular frequencies of noise is possible by linear filtering, which may be performed in part electronically. Another possibility (which is, contrary to linear filtering, irreversible in that the original signal cannot be reconstructed from the processed one) is piecewise-linear approximation, with a prescribed bound on the maximum error.

The amount of recorded data may be significantly reduced, without any loss of information, by piecewise-linear approximation and other means. In the case of the ECG, for example, the number of points in one record may be very large and most points redundant since long sections are essentially straight lines. A part of the record can thus be deleted. This not only helps to reduce memory requirements but may also simplify further processing.

There are other types of preprocessing and their purpose often is to transform the original signal into a different form, more suitable for further processing. An example is the calculation of the power spectrum of a waveform, used in some programs for ECG analysis.

Another segment of the preprocessing part of the program may deal with pooling of data from several complexes within the recorded signal. This may be advantageous especially in the case of signals with a repetitive character, such as the signals generated by the almost periodic pumping action of the heart. Pooling typically takes the form of averaging of either entire complexes or the chosen features themselves (amplitudes, durations etc.). It has been argued that since these cycles are often not fully repetitive (for example breathing has a significant effect on the circulatory phenomena) a better approach would be to extract a few complexes typical for the record rather than add all complexes indiscriminately together.

After passing though the preprocessing stage of the program, the signal has a form suitable for the identification of features chosen at stage 1: In the processing of the ECG these features may be the amplitudes and timing of the P, Q, R, S and T waves. Subroutines for reliable recognition of these features in the digitized signal, tailored to the given application have then to be written and tested, to guarantee satisfactory performance on even the most pathological cases (from the programmer's point of view); the program must be able to handle signals spoiled by noise or artifacts, waveforms of unusual configurations etc. It has to be remembered that a mistake made by the program at any stage is carried through all the rest of processing and errors in general cumulate.

Recognition of features is the first part of the processing program (and also one which is subject to most randomness and abnormal input). For this reason, the program should have the option to reject data which do not satisfy certain minimal conditions; so, for example, if the P wave in an ECG is not well defined, it may be better to leave it (and possibly the whole complex) undefined, rather than to make a guess. Since the signal is in a number of cases affected by artifacts due to the motion of the subject, irregular breathing etc., the program also includes a portion designed to eliminate sections of the record affected in this way. In the case of the ballistocardiogram, sections of the record may be completely spoiled by sudden jumps of the amplitude of acceleration. These jumps may be caused by collisions between the ECG bed and the edges of the air bearings. They can be recognized and the affected sections of the record eliminated from further processing.

2. Classification

On the basis of the chosen and located feafures and an accepted statistical criterion, the sample is now classified as belonging into one of the considered classes. So an ECG record may be classified as normal or abnormal on the basis of the timing and amplitudes of the P, Q, R, S and T waves. The classification may, however, sometimes be made only with a limited degree of likelihood, when the given record represents a "borderline case". If the probability of correct assignement of a given signal (sample) is not sufficiently large, we may desire not to classify the signal at all. A number of decision schemes have been developed for classification and some will be described later.

3. Testing the Program

When the program is completed, it is tested upon samples with known classification (the "test set"). Statistical theory offers several methods for the choice of samples for the test set. The usual approach is to sacrifice a part of classified samples for this purpose (that is these samples are used only for testing and not for the design of the rule) since testing of the program on the design set may result in a misleading estimate of its performance. If the performance of the program is satisfactory, the design of the program is finished, otherwise we have to return to stage 1, profiting this time from the experience gained during the first design.

When the program is completed and tested, it should be certified as suitable for practical use in clinical situations. The problem of standards, testing and related questions have been discussed, for example, by Caceres and Dreifus [2]. It has been suggested that a library of classified records representative of the conditions that can be encountered in practical situations should be established and quantitative criteria for a minimum acceptable performance agreed upon. In the case of the ECG some of these criteria would, for example, be: the amount of noise present in the signal that the program should tolerate (give correct diagnosis), a limit of precision of measurement that the program can work with, the maximum acceptable proportion of incorrect results in the identification of QRS complexes. Only those programs which pass this test should be recognized as suitable for practical use.

The described scheme of a typical approach is illustrated by the block diagram in Fig. 1. We will now discuss the individual stages in more detail.

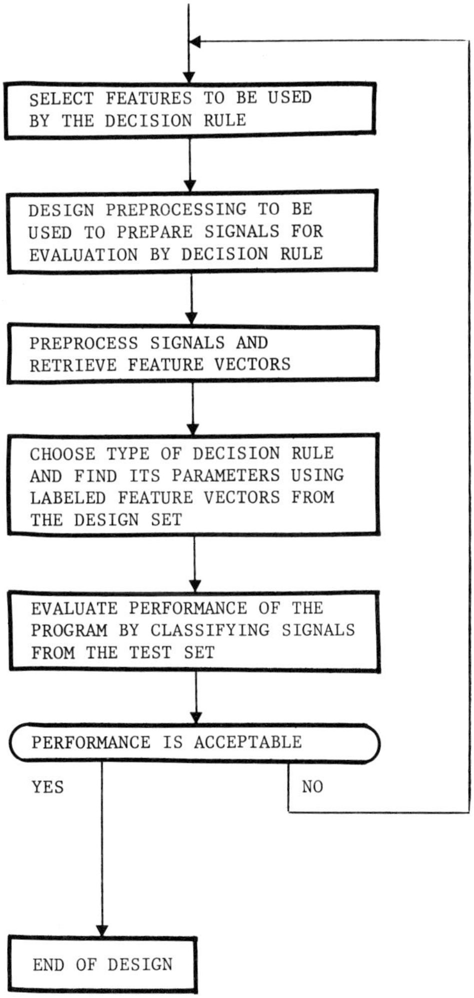

Fig. 1 A flowchart of the design of an algorithm for waveform interpretation.

B. Feature Selection

This step has so far always been performed by humans, sometimes with the help of computers (an example is described in section III. B). In an ideal situation in which the origin of the waveform can be described by a certain known number of easily obtainable parameters, it is, of course, ideal to use these parameters. So, for example, if the waveform is generated by a second order linear differential equation, we can calculate parameters which fully characterize the recorded waveform. Such a situation is, unfortunately, practically inexistant. Nevertheless even a partial understanding of the physical meaning of the waveform is helpful for an intelligent choice of features and should always be sought.

In the more usual situation the human operator compares the available classified records and tries to find features which discriminate efficiently between the classes he knows to be present. This process is largely intuitive and not well understood. As a tool in this process the computer may be used to confirm or disprove the operator's feeling that a certain feature is typical for a given class (or its part). Information gathered by previous medical research is most helpful in this process. The successful ECG classification programs are an example of programs mostly based upon the accumulated knowledge gained over the decades of ECG investigation.

This heuristic approach is not ideal for several reasons. First of all, it requires a lot of time on the part of highly qualified researchers, a time which could be spent in a more creative way. Secondly, there is no guarantee that the features chosen by human operators are the most effective ones. It may be that other features exist, which differentiate between the given classes more accurately; their use with a given decision rule may result in a smaller proportion of erroneous decisions.

It would be ideal if there were an algorithm which would (possibly with some help from a human operator) accept a set of classified records, process them and output a set of features that guarantee acceptable classification. Such an algorithm would be useful not only for the purpose of automatic classification but also for the understanding and "manual" application of the method. Suppose, for example, the ECG has just been invented and a certain number of records collected from subjects in known conditions (for example normal and abnormal records). With the help of our algorithm we could, in a very short time, find features typical for records from any given class (and also subdivide our classes into subclasses according to differences within a class). The algorithm would, of course, be aiming at those features which are typical for a given class of records and distinguish it, clearly enough, from the other classes. These features could be used either for computer classification or for "manual" evaluation of records by medical personnel. This would save decades of human work and make the method practically useful much sooner than under the present conditions. There is no doubt that it will take a long time before such a general method is even approached. On the other hand, waveform recognition is probably the one area among many pattern recognition applications which offers itself best for such a solution.

C. Preprocessing

We will only discuss three methods of preprocessing: linear filtering, nonlinear piecewise linear approximation and clustering of repetitive complexes into groups.

Linear filtering will be described without much detail, since it is the subject of a number of publications (read for example Gold and Rader (1969)). It is a highly developed field and its description is beyond the scope of this chapter. Its widespread use requires, however, that at least the basic concepts be mentioned for completeness. This is provided later in this section.

Piecewise-linear approximation will be described because it is a relatively new method which so far has not been applied in practical situations. It is hoped that a section on this method will contribute to its deeper study and an increasing amount of applications.

The same reason lead to the incorporation of clustering of similar complexes, which is relatively new and has, in the opinion of the author, a great potential both as a method of data compression and a tool, simplifying classification. No practical use of it has, however, yet been reported in the literature.

Other methods of preprocessing have been described in the literature and occasionally used in isolated cases but their use is not widespread, usually because of the computational requirements they pose. Some methods have widespread use but not in the processing of cardiological data (this is the case with power spectra) and will not be discussed either.

In our outline, we have included algorithmic recognition of features into preprocessing. This topic will not be discussed in detail since there are no generally applicable methods for this problem. A number of algorithms for specific problems have been described in the literature (the journal Computers and Biomedical Research is a good source of information on this subject).

1. Linear Filtering

This type of preprocessing is used almost universally. The reason is that linear filtering is ideal for the elimination of undesirable components characterized by a limited frequency range of occurrence, separated from that of the useful signal. Noise of this type forms the major part of undesirable signals (the rest being in general unseparable artifacts). We will call the undesirable component "noise", although this meaning is often somewhat different from the usual sense of the word. Much of the noise is caused by stray fields affecting the electronic circuits, improperly placed electrodes, building vibrations etc. Some of it, however originates from the observed system – the subject's body itself. This is the case with the effect of respiration on the circulatory phenomena, the muscle tremor and those components of the useful signal itself which, by their presence, complicate the analysis of the record.

Linear filters are characterized mainly by their ability to modify the frequency characteristic of the signal. This property may be used to differentiate or integrate the original signal. Its main application is, however, to attenuate those undesirable components of the signal which have their harmonic components in a certain frequency range (from this the name filter is derived). Digital linear filters are the digital counterparts of electrical and mechanical filters and their design is guided by similar principles. According to their behavior they are often described as low-pass, high-pass or band-pass filters – depending upon the frequency range they let through relatively unaffected. The basic types of linear digital filters are

i) filters working in the time domain based on the linear operations of addition, subtraction and multiplication by constants performed on original signals and

ii) filters working in the frequency domain working with harmonic components of original signal modified by multiplication by complex valued factors. (Transformation of the original signal and an inverse conversion into the time domain is necessary but can be performed very efficiently by means of the Fast Fourier Transform. In some cases analysis is performed on the transformed rather than the original signal and inverse conversion is then not neccessary.)

The general form of the expression for a time domain digital filter is

$$y(n) = \sum_{r=1}^{p} u(r)y(n - r) + \sum_{r=m}^{q} v(r)x(n - r) \tag{1}$$

Here $y(i)$ is the value of the filtered signal (the output of the filter) at time $t = t_0 + i\Delta$, $x(i)$ is the value of the input (original) signal. The value of p determines the order of the filter. Filters of type (1) are called recursive filters. When the term $\Sigma u(r)y(n - r)$ is missing, the filter is called non-recursive (it does not use its own previous outputs to calculate its current output). The "moving average" filter described by

$$y(n) = \sum_{r=p}^{p} w(r) \times (n - r) \tag{2}$$

also belongs to the family of linear digital filters.

The operation of filtering is often referred to as smoothing. Linearity is a very important property of these filters. Another useful characteristic of linear filters is the reversibility of the filtering process, whereby the original signal can be recovered from the filtered signal. This is evident from expression (1) above, whereby the input can be calculated from the output using a formula obtained by solving a set of difference equations derived from (1). When correctly designed (the design has to take into consideration the frequency characteristic of the original signal and its noise components) in managable situations (when the noise and useful signal are mostly contained in non-overlapping frequency ranges) linear filters can give spectacular results as is shown in Fig. 2. There is, however, a cost that we have to pay for their filtering effect: all filters generate a distortion of the original signal both in amplitude and timing (Fig. 3). This distortion adversely affects the pattern recognition stage later.

The extent of distortion depends on two factors: the frequency components of the useful signal and noise, and the design of the filter.

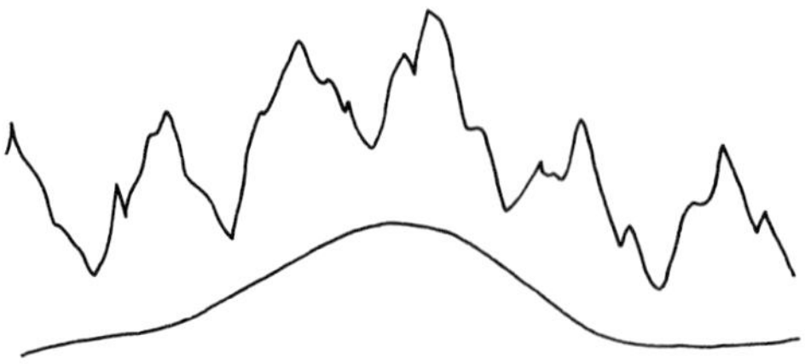

Fig. 2 An ECG P wave before and after linear filtering [27]. Note that the high frequency noise is completely eliminated but the low frequency component very little affected.

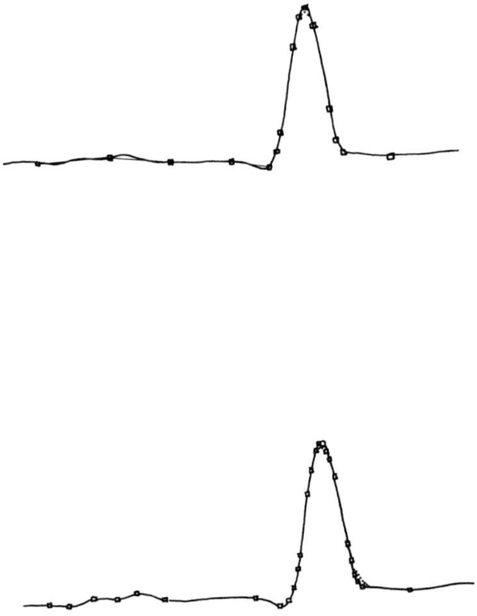

Fig. 3 Application of PLEA (—, —) and a linear filter (---) to a signal partially processed by a linear filter (——). a) $\varepsilon = 50\ \mu V$; b) $\varepsilon = 25\ \mu V$. (From Tomek [31]).

The first factor is beyond our control, the second can be influenced by choosing values of parameters and order of filter. The higher the order of filter, the greater freedom of affecting the frequency components of the signal and in general better filtering. Higher order filters require, however, more terms and thus more processing time. A compromise has to be made between these conflicting trends.

Comparison of digital and analog filters shows that digital filters are truly linear, to the extent of digital accuracy, while analog filters are always only approximately linear since such are their constituent elements. Digital filters are also stable (analog filters are affected by the environment and aging; their performance is thus subject to changes). The main disadvantage of digital filters, in comparison with the analog ones, is that they require some computer time at the expense of other processing. This is important especially in on-line applications where processing has to follow recording continuously and with little delay.

2. Piecewise-Linear Approximation with a Bound on Error

It has been mentioned that linear filters always introduce some distortion. The unpleasant fact is that this distortion depends upon the shape of the signal and is thus variable and unpredictable to the extent to which the possible shapes of signals are unknown in advance. Distortion affects both amplitude and timing (Fig. 3). Piecewise-linear approximation with a prespecified bound on error (PLEA for piecewise-linear epsilon approximation, epsilon being the bound) is a method which removes completely the disadvantage of the unpredictability of distortion and reduces the distortion in timing. We are going to describe it in more detail since it is relatively unknown.

Intensive research into efficient algorithms for PLEA implementation motivated by waveform recognition started relatively recently. By now a very efficient numerical algorithm has been found (Tomek 1975 a); its performance is, in terms of CPU time, at the same level as that of linear digital filters. This comparison is only approximate since the speed of digital filters is variable and decreases with increasing order of filter and PLEA depends on the value of epsilon and the shape of the original signal. The use of piecewise linear approximation has been advocated by researchers in pattern recognition for some time (Pavlidis, 1967). The main problem with its use, apart from its novelty, has been the limited speed of the existing numerical algorithms. Now that PLEA can be performed very fast, there are no obstacles to the use of PLEA in practical applications.

A typical PLEA algorithm works as follows (see also Fig. 4): The starting point x(1) is

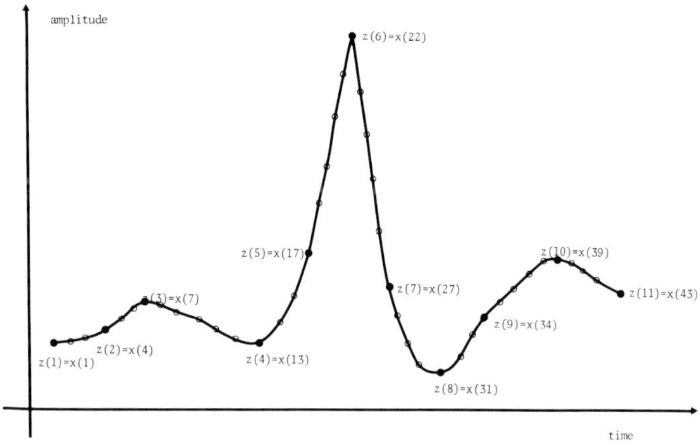

Fig. 4 The principle of the maximum length segment PLEA described in the text.

accepted as the beginning of the first interval: z(1). Successive points are then examined as canditates for endpoints of the first interval i.e., if the straight line between z(1) and x(i) does not depart by more than ε from any point x(j) $j = 2, 3, \ldots, i - 1$ in amplitude, then x(i) can be the endpoint of the first interval. The candidate endpoint with the maximum index (the longest interval) is accepted as the end point of the first interval: z(2). This point is also the beginning of the second interval whose endpoint is found in a similar way. This search continues as long as there are any points left. A comparison of the application of a linear filter and PLEA to a signal previously filtered by an analog filter is in Fig. 3.

The comparison of PLEA (which acts as a non-linear digital filter) and linear digital filters shows that the main advantages of linear filters are the reversibility of their operation and their ability to eliminate noise whose frequency components are in a range separated from that of the useful signal. The relative advantages of PLEA are the specifiable bound on the maximum allowable amplitude distortion, significant data reduction (which depends upon the type of the signal and the prescribed maximum allowable error – a typical saving is 60%) and its power to detect certain types of important geometric features (such as inflection points and peaks of waves whose amplitude exceeds ε). It seems that, in most cases, the best approach is a combination of a linear filter (to remove some of the noise) and PLEA, which reduces the amount of data by constructing an approximation with known accuracy and simplifies location of features in the second stage of processing.

As far as this author knows, PLEA has not yet been applied in practical situations. It can be expected that its application will encounter some resistance for a number of reasons: the wide acceptance of linear filtering and the neccessity to reformulate existing programs not only technically but also conceptually, which the introduction of PLEA would require. It seems, however, that the advantages are so significant that this will eventually be done in cases where PLEA can be expected to be more powerful than simple linear digital filtering.

We have seen that extraction of features is one of the first stages of the processing of a signal. To demonstrate that PLEA processing is a suitable method of preprocessing, the question, how well this process can be performed with data in this new form, has to be answered. To test the suitability of piecewise-linear approximation for feature extraction from the point of view of speed and accuracy, the effect of PLEA processing on feature extraction has been tested on the recognition of the F, P, I, D points on brachial pulse wave recordings (Tomek 1976 c). It has been found that the time required for identification of these points on PLEA data is on the average about one third of the time required on the original data. The accuracy of identification is practically identical.

3. Extraction of Typical Complexes

Biomedical signals very often consist of almost periodically repeated complexes. This is true particularly about signals generated by the almost periodic action of the heart – the ECG, PCG, ACG, BCG etc. In many cases, the individual complexes can be considered as "pure" complexes corrupted by noise of internal (such as the effect of breathing) or external (the effect of amplification, filtering, recording devices etc.) origin. Since the noise is random both in amplitude and in time of occurence, averaging by summation of individual complexes can be expected to eliminate much of the unwanted components. In fact, special devices known as averaging computers are available to do this as a part of the interface between the measuring device and the computer. The larger the number of averaged complexes the closer the resulting averaged signal will be to the "pure" one. This approach simplifies evaluation and improves its accuracy, since errors in identification of features become less frequent. At the same time, the amount of data to be stored can be significantly reduced. For these reasons, averaging has become a popular method in the processing of biomedical waveforms.

The justification of averaging given above is quite valid in those cases, where individual complexes can be thought of as distorted variations of a single "pure" form. This is, however, very often not really the case since in many records some complexes are generated under quite different conditions and the resulting difference in shape is not simply due to noise. For example, irregularities of heart rate cause significant differences between individual heartbeats and these appear as quite different BCG complexes. Since this case is more usual in abnormal subjects and since these are the subjects which are particularly important in medical applications, this fact constitutes a serious objection against averaging. Another objection to averaging is that by elimination of differences between complexes we may lose valuable information. So, for example, BCG complexes vary continually with breathing. It is of interest to know, what shape the complex has for maximum inspiration and minimum inspiration.

The above two objections apply to a broad field of applications in which averaging is undesirable. In the first case, more or less randomly assembled complexes can be averaged to generate a complex which does not resemble any of the constituent ones, in the second case, much valuable information may be discarded if the whole record is replaced by an average complex.

It has been proposed (Previtt 1972, Tomek 1972) that a more natural method of elimination of redundant information and selection of the most representative data is to extract only the "typical" complexes. (This is, in fact, the approach taken automatically by human evaluators.) These are the complexes that appear relatively often and/or are sufficiently distinct from other complexes in the record. Complexes should thus be grouped into "clusters" of similar complexes, and each cluster represented by a complex typical for the given group (for example the average of the cluster). This formulation puts the problem into the domain of pattern recognition (clustering). It is known, however, that clustering has its problems:

a) There is the problem of the meaning of similarity:

We have to define similarity numerically. It is customary to have each complex represented by a certain number of coordinates (as a vector). Complexes are similar if the distance of the representing vectors is small (often meaning smaller than a threshold), dissimilar otherwise. The difficulty is, how to choose the coordinates and the definition of distance so that the relationships of similarity generated by these two choices are the same as those that we subjectively require. A possible choice of distance is the euclidean distance; if $c = (x_1, x_2, \ldots, x_n)$ and $d = (y_1, y_2, \ldots, y_n)$ are two vectors, the euclidean distance is

$$\text{dist}\,(c, d) = \sqrt{\sum_{i=1}^{n} (x_i - y_i)^2} \tag{3}$$

Another common choice is the absolute value distance:

$$\text{dist}\,(c, d) = \sum_{i=1}^{n} |x_i - y_i| \tag{4}$$

These are the most obvious and most often used definitions. There are others which take care of the shifting baseline or changing scale etc. They are in general more difficult to evaluate numerically.

The problem of a good choice of coordinates is probably more difficult, since it is not known how humans percept differences in shape, and consequently it is not known how to describe shape quantitatively (most of the work related to this problem has been done in the context of picture processing and is dealt with in the book by Duda and Hart, 1973). The most straightforward approach is to identify the representing vectors with the numerical values of the complex's amplitudes. It turns out (Fig. 5) that this representation often gives results which are in variance with human perception. For example, a small difference in amplitude over a long stretch of time causes the numerical value of distance to be large even though the complexes may appear to be almost identical to humans.

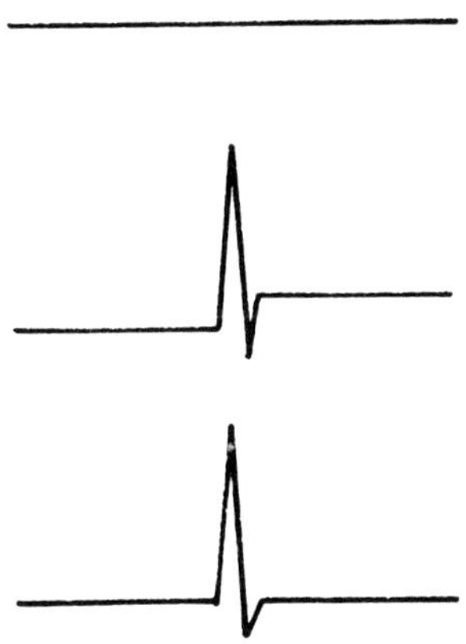

Fig. 5 The pointwise eucleidian distance between the uppermost and lowermost waveforms is smaller than all other distances (from Tomek [29]). Yet the two lower signals are clearly more similar to one another.

Another possibility is to perform the Fourier transform on the original coordinates and use the obtained Fourier coefficients as representatives of the complexes. This is tempting since Fourier coefficients are generally considered to be good shape descriptors. Unfortunately, this is not always so. Fig. 6 shows what happens for example when one coordinate is significantly different and others are almost identical; the resulting shapes look quite different but their distance is relatively (in comparison with distances to a third complex) small.

It has also been proposed to use, as coordinates, amplitudes of points which are likely to be important for the shape of the complex (Tomek 1973): points in which an extremum of the (suitably defined) curvature is attained have been used in picture processing (where this was

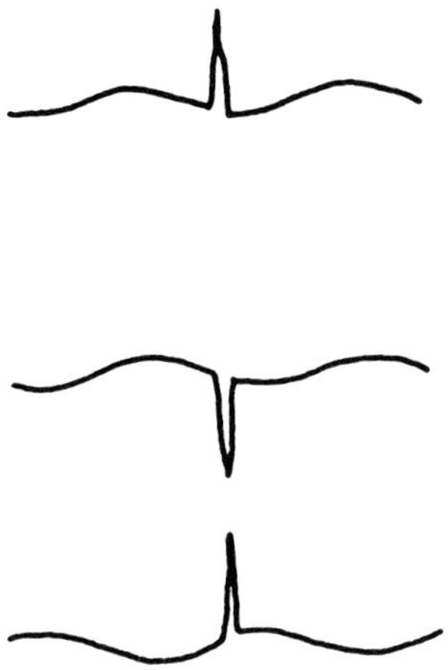

Fig. 6 Eucleidian distance (with Fourier Coefficients) between the two uppermost waveforms is smaller than all other distances (from Tomek [29]). Again this is in conflict with our perception of the uppermost and the lowermost signals being similar.

motivated by psychological experiments). This suggests that perhaps the local extrema of amplitude and curvature are the determining factors in human perception of shape and should thus be used for the representation of complexes.

The best choice of coordinates seems to be as follows. If we know what are the important features of our problem, then these should be used as the components of the representative vectors. Let us consider ECG processing, assuming that the PQ time and QR amplitudes are the most important parameters in the considered context. Each complex should then be represented by a vector with two coordinates (PQ time, QR amplitude) and similarity of complexes measured by the distance of individual vectors. To complete the definition of our problem we have to solve the problem of scaling. The contributions of differences of individual coordinates to the total distance have to be multiplied by weighting coefficients to allow for their different importance and scale. The PQ time is in units of time: should it be measured in seconds, miliseconds? Should the amplitude of the QR wave be measured in volts or milivolts? The scale will determine the relative importance of individual coordinates.

If we use a threshold in our process the largely subjective choice of its value is a similar problem.
b) The second difficulty is that clustering is not an unambiguously defined problem.

This is because clustering is derived from a human activity which is not itself unambiguous. Human subjects will often hesitate how to cluster objects into separated groups. It is not surprising that the numerous clustering methods (often developed for specific problems) do not give the same results in identical situations (Fig. 7). Fortunately the number of known methods is large and comparison allows the user to choose one which gives results in accordance with his needs. An example of the application of the method of typical complexes to BCG patterns is shown in Fig. 8.

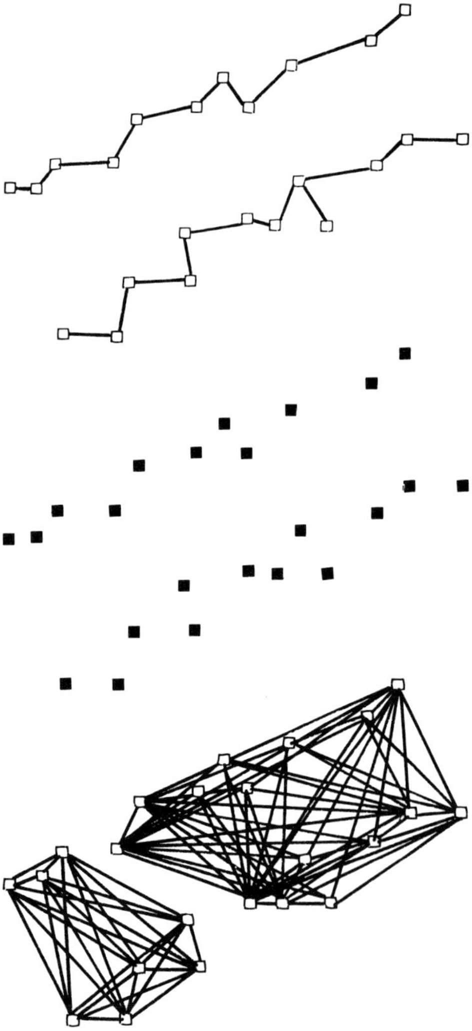

Fig. 7 Clustering of a set a) obtained by the nearest neighbor clustering algorithm (b) and by the furthest neighbor algorithm (c) (from Duda, Hart [6]). This example shows how different results different algorithms can give.

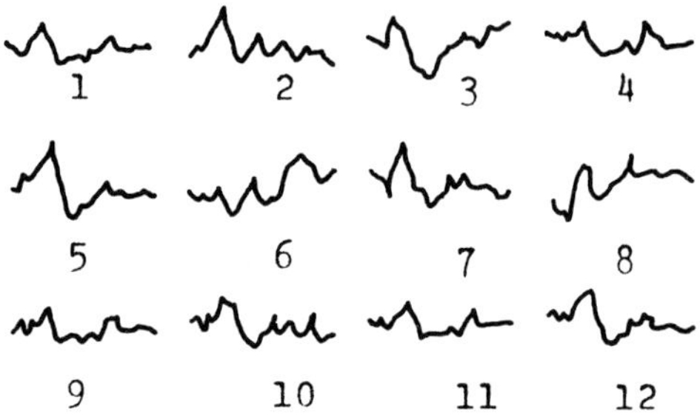

Fig. 8 An example of clustering of BCG complexes. Complexes (1,4,9) are represented by pattern 11, (5,7,10) by pattern 12 (from Tomek [28]). The set of complexes 1–10 can be meaningfully replaced by complexes 11 and 12.

The clustering method discussed in this section is relatively new and has some problems (described above). Its practical use has not yet been reported. Its potentials seem to be large: data reduction without loss of information, enhancement of information hidden in the record by local irregularities. Finally if we really feel that human performance is at least a provisional model of ideal waveform recognition programs, then this typically human feature must be incorporated into their structure.

D. Classification

This section takes us from the rather ill-defined problems related to human perception of shape and its processing into the domain of mathematical statistics, in particular its part concerned with classification. We will have to use a few basic concepts from statistics. A qualitative outline of the necessary terms is given in Appendix 1.

Let us assume that we are processing samples represented by n-dimentional vectors, such as $x = (x_1, x_2, \ldots, x_n)$. These samples are taken from a space X. Let our samples be generated by a process dependent upon a certain number of parameters. In the case of the ECG, our samples are vectors representing ECG records, the generating process is the electrical activity of the heart plus the interfacing effects of instrumentation, the parameters are given by the subject's condition and external influence of measuring and recording devices (for example filters and amplifiers). Let one of these parameters be discrete, assuming values 1, 2, . . . C, representing classes into which the generating process can be classified. In the ECG example, one group might consist of normal subjects, the other of abnormal ones (we thus have $C = 2$). This will be assumed throughout the remainder of this section. This assumption considerably simplifies presentation of the main results, but can be easily removed and the methods reformulated. Consideration of the general case is, however, somewhat confusing in other situations, mainly for the number of auxiliary indices by which the relevant expressions have to be appended. (The interested reader is urged to consult any of the mentioned books on pattern recognition for more details.) In biomedical applications this situation is often generated artificially; the classes are not given "naturally" but are constructed according to clusters observable in the distribution of recorded samples.

Some of the parameters of the generating process are random, their value is subject to random fluctuations depending in general upon the value of the discrete parameter. They can thus be described only statistically. We will assume that they can be described by probability density functions associated with the values of the discrete variable. Returning again to our ECG example, the ECG record depends upon the exact physiological condition of the subject and external factors; there are differences between different records obtained from the same subject, as for example, the time of the day, the way the electrodes are applied, the condition of the measuring device etc. affect the record. Then there are differences between individual normal subjects. Given a sample generated by this process, it is in general impossible to decide with certainty which value of the discrete parameter was present during the generation (in other words to which class the sample belongs). Always a certain overlap is possible, there are, for example, abnormal subjects with practically normal ECGs and vice versa. Any specific judgment will thus have to be based upon statistical evidence and possibly have statistical formulation, such as, that it is more likely that the sample was generated with discrete parameter value 1 rather than 2: The probability of the subject being normal is P(1), the probability of abnormality being P(2).

The above qualitative description can be formalized as follows: Let $x \in X$. Denote the probability density (characterizing the relative number of samples from the given group in the given part of sample space X) associated with class C_i, $i = 1, 2$, by $P_i(x)$; the information contained in $P_i(x)$ is not sufficient; we also have to know in what proportions samples from the two classes occur. (If we are examining students of physical education, relatively very few will have abnormal hearts and a somewhat abnormal ECG is not as strong an indication of a diseased heart as if we were examining subjects complaining about other suspicious symptoms.) This information is contained in the a priori probabilities p_i, $i = 1, 2$. (The name is derived from the fact that they are probabilities of a sample belonging to a group, assigned before the sample is closely examined). To be consistent with our approach to samples as being generated by a process governed by values of certain parameters, we can say that a priori probabilities are probabilities associated with our discrete parameter, while probability densities are associated with the remaining parameters (the value of the discrete parameter being constant for a given density function).

Now that we have all the necessary information, we will describe how to use it to set up a numerical rule for assigning samples to one of the two groups on the basis of the value of p_i and $p_i(x)$. As we have mentioned, our groups in general overlap (meaning that there are regions in which both $p_i(x)$ and $p_2(x)$ are nonzero and so occurence of samples from both groups is possible), so that we cannot expect to be able to design a rule which will not make incorrect assignments. In general, we will thus have to be satisfied if, on the average, our rule will make as few errors as possible. This requirement is known as the minimum expected error criterion and we will show how it can be used to derive a numerical rule for classification.

1. The Bayes Rule

Let $R(x)$ be the rule which to any $x \in X$ assigns a number: the number of the class to which x is assigned. In the case of two classes, $R(x)$ can therefore have either value 1 or 2. The decision rule thus partitions X into two subspaces X_1 and X_2 (every point of X belongs into exactly one of the two subspaces since it is assigned either to class 1 or to class 2). We want to minimize the average error over a large number of performed assignements. In statistical terminology, this is the expected error. The expected error for a given rule (expressed in terms of the function R or in terms of subspaces X_1 and X_2) is

$$E(R) = E(X_1, X_2) = E(x \in Ci, R(x) \neq i) \tag{5}$$

The right hand side is the relative proportion of wrong decisions (which occur whenever a sample belonging to a class Ci is assigned to class $Cj \neq Ci$). The error is averaged over the whole space X. For a function $f(x,i)$ defined on X and a set of discrete values i (representing, in our case, the two groups) the expected value is

$$E(f) = \int_X \left\{ f(x,1)p_1 p_1(x) + f(x,2)p_2 p_2(x) \right\} dx \tag{6}$$

In our situation, the function $f(x,i)$ counts errors and so its value is 1 if an error is made ($R(x) \neq C(x)$, rule does not assign x to the proper class), and 0 if x is assigned correctly ($R(x) = C(x)$). When this function is substituted into the above expression we obtain

$$E(X_1,X_2) = E(R) = \int_{X_1} p_2 p_2(x)dx + \int_{X_2} p_1 p_1(x)dx \tag{7}$$

Since p_i and $p_i(x)$ $i = 1, 2$ are given, we can minimize $E(X_1,X_2)$ only by a proper choice of X_1 and X_2. (This is indicated notationally by having X_1 and X_2 be arguments of E.) Let us now take an arbitrary point x from X. The "contribution" of point x to the value of $E(X_1,X_2)$ is $p_2p_2(x)$, if x is in region X_1. This is because if a sample from class C1 happens to have value x, its contribution to $E(X_1,X_2)$ is zero (since it will be classified correctly). If, however, x is the value of a sample from C2 the contribution is $p_2p_2(x)$ since the sample is then classified incorrectly (into class C1 by definition). In order to minimize the contribution we thus have to use the following rule (see also Fig. 9):

$$R(x) = \frac{1}{2} \text{ if } p_1p_1(x) \gtrless p_2p_2(x) \tag{8}$$

Application of this rule guarantees that the argument of our integral has always the minimum possible value and thus the value of the integral is minimum as well. Let us repeat that this does not mean that we will not make errors but rather that when we assign a large number of samples using this rule, the number of errors made will be smaller than that made by any other rule.

The rule just derived (known as the Bayes rule) is only one of a class of rules which can be derived in a very similar way. The derivation is based upon a criterion function (a "cost"

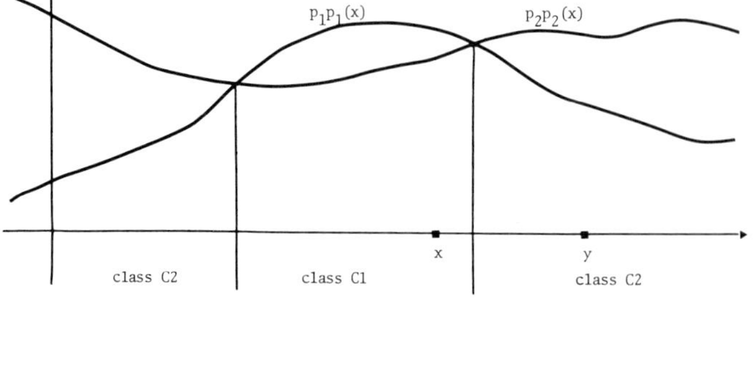

Fig. 9 The minimum expected error decision rule. The relative value of $P_i P_i(x)$ decides upon the classification: sample x would be assigned to class 1, sample y to class 2.

function) the expected value of which is to be minimized by a proper choice of X_1 and X_2. This more general "minimum expected risk" (MER) criterion is defined as follows: Let each decision be associated with a fixed cost – a cost for a correct decision and a cost for the wrong decision. The objective is to minimize the expected cost. The minimum expected error is a special case of this rule, one in which the cost of an incorrect decision is 1 and that of a correct decision is 0.

The above described most often used criteria result in decision rules which have the same form; they consits of comparison of probabilities just as in the formula derived above. The difference is that, in the general case, the probabilities in the above expression are multiplied by a constant factor, which leads to a shift of the boundary between X_1 and X_2. Numerically this is expressed by replacing terms $p_1p_1(x)$ and $p_2p_2(x)$ by $c_1p_1p_1p_1(x)$ and $c_2p_2p_2(x)$. Coefficients c_1 and c_2 depend upon the values of costs of assigning a sample, correctly or incorrectly, to class C1 or C2. The difficulty associated with the use of the MER criterion is, that in most cases there is no objective basis for the choice of specific numerical values of 0_1 and c_2.

It is evident that in all cases, ambiguous situations arise whenever instead of inequality, the comparison results in equality. It is easy to see from the meaning of the rule that in these cases the most natural outcome of the rule is not to classify the sample at all. In all practical situations, the occurence of such a case is practically impossible since the intersection of the two functions (i.e. equality) usually defines a region (decision boundary) of zero measure and thus zero probability – meaning that such situations are extremely unlikely to occur. The alternative of equality has, of course, to be considered in the program.

We have mentioned that if the probability of correct assignement (diagnosis) is not sufficiently large it may be desirable not to classify a sample at all. This type of classification is known as classification with a reject option and will be described later.

In the multiclass problem, the Bayes decision rule is essentially identical: x is assigned to that class which gives the maximum value of $p_ip_i(x)$. The derivation of the rule is the same as that for the two-dimensional case.

We now have an optimal rule. This, however, does not quite solve the problem. In order to be able to apply it we have to know p_i and $p_i(x)$, $i = 1, 2$. These are never known accurately. We can only approximate them by using our knowledge of the distribution of the collected and classified samples. This can be done in a number of ways. We will describe some of the most popular approaches and disscuss their relative merits. For a more detailed treatment of most of the presented material, see for example Duda, Hart (1973), Fukunaga (1972) or Meisel (1972).

1.1 Use of representatives

This is the most straightforward but also the least effective method. It is not strictly based upon the optimal rule but can be interpreted as such in certain special cases (it will be seen that this is true for some of the other described methods too). It works as follows:

Let $D_1 = (x_{11}, x_{12}, \ldots, x_{1q})$ and $D_2 = (x_{21}, x_{22}, \ldots, x_{2q})$ be two clusters of *samples* (Fig. 10). Let us find the average of set D_2 (samples from class C2):

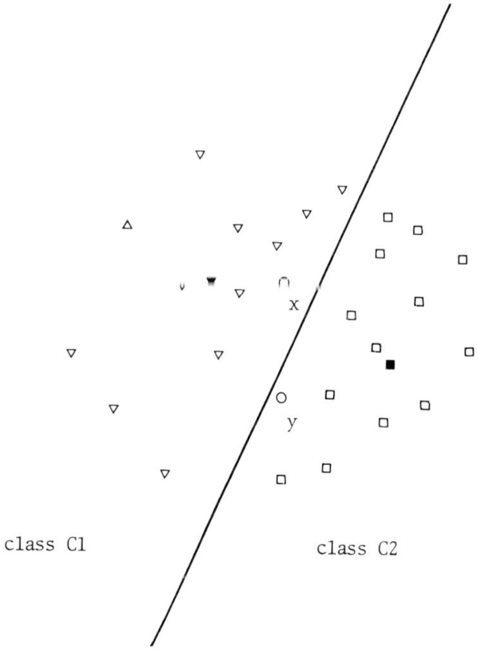

Fig. 10 The method of representatives. Representatives (means) of clusters are denoted by full symbols. The decision boundary is shown by the full line. Sample x is classified into class 1, sample y into class 2.

$$x_{2a} = \frac{1}{q} \sum_{i=1}^{q} x_{2i} \tag{9}$$

and similarly x_{1a}. To classify a new sample y, we calculate its distance to both averages (centroids) and assign y to that class which is closer (as measured by the distance to centroids). The method can also be used for more than two clusters. This simple rule is useful only in the simplest cases (Fig. 11). Its advantage is the computational efficiency; we only have to store the centroids and calculate only a few distances to classify a sample; in fact distances do not have to be calculated since

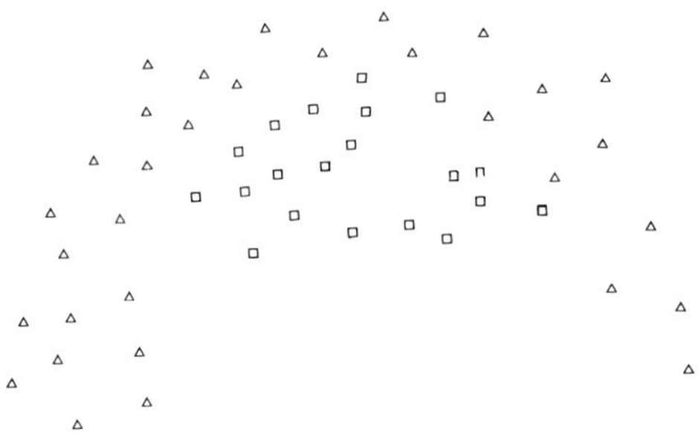

Fig. 11 A situation in which the method of representatives cannot be used: representatives would generate a meanigless boundary.

$$\text{dist}\,(y,x_{1a}) \lesseqgtr \text{dist}\,(y,x_{2a})$$

is the same as

$$2 \sum_{i=1}^{q} y_i(x_{2ai} - x_{1ai})$$

$$\lesseqgtr \sum_{i=1}^{q} (x_{2ai}^2 - x_{1ai}^2) \tag{10}$$

so that only the calculation of linear terms is necessary.

1.2 Methods based upon approximations of density functions

These methods are the only ones that use explicitely the optimum rule. The a priori probabilities, needed in (5), are approximated by relative frequencies:

$$p_i = \frac{N_i}{N_1 + N_2} \tag{11}$$

(here N_i is the number of samples in D_i). Approximation of $p(x)$ is performed by assuming a specific type of function with unknown parameters and finding optimal (in one of a number of senses) values of these parameters. So, for example, we assume that

$$p_i(x) = \sum_{j=1}^{k} \alpha_{ji} \frac{1}{\sqrt{2\pi}\,\sigma_{ji}} l^{-\frac{1}{2}\left(\frac{x-\mu_i}{\sigma_i}\right)^2} \tag{12}$$

and estimate parameters α, σ, μ. The general case described by equation (12) is too complicated to be disscussed here in any detail. Let us mention that in the case of $k = 1$ (the "unimodal" case) there is no need to consider $\alpha \cdot \mu$ is estimated as the mean value of sample coordinates

$$\mu \doteq \frac{1}{p} \sum_{j=1}^{r} x(j) \tag{12a}$$

The value of σ is obtained from

$$\sigma^2 \doteq \frac{1}{p} \sum_{j=1}^{r} (x(j) - \mu)^2 \tag{12b}$$

In quite general situations, approximation of density functions is based upon standard methods of functional approximation. Some authors use the statistical context to formulate methods particularly suitable for probabilistic use. More information can be found in Meisel (1972).

The advantage of these methods based on the optimal decision rule is that they can handle even the most unusual types of distributions, at least theoretically. In practice this may be difficult and time consuming. The assumed form of the function limits the distributions that can be reasonably well approximated. Also the estimation of the number of modes (Fig. 12) of the distribution approximated by (12) is difficult.

The representativeness of the design sets D_1 and D_2 is essential for the good performance (by representativeness we mean that samples are distributed in a way typical for the given problem; in other words the relative occurence of samples in X is related to the numerical values of probabilities). If the design sets are not representative, the approximation of probability densities and a priori probabilities will be poor and with it the performance of the method. This is, of course, true for other methods as well.

Methods using approximations of probabilities and the optimal rule are in general quite inefficient computationally. For the method to work well the approximation has to be good.

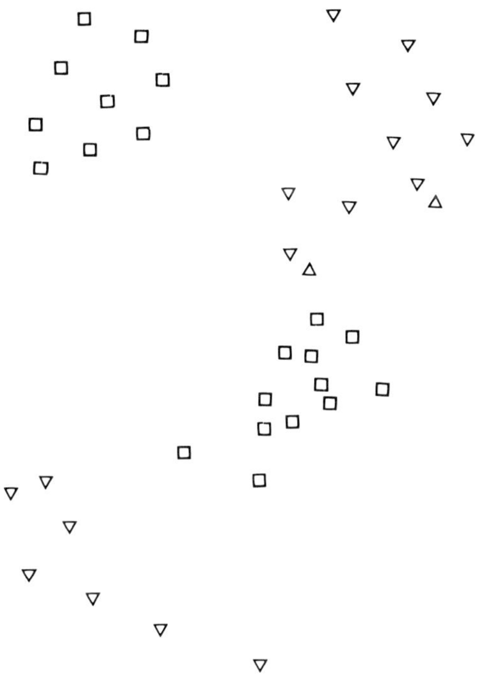

Fig. 12 Two multimodal distributions. Samples are grouped in clusters, more or less mutually separated.

This in turn often requires a large number of parameters. These have to be stored permanently and used for calculation of the function for every new sample. Since the functions used for approximation are usually rather complex, the evaluation is very time consuming. Another disadvantage of this approach is that, although with large design sets and sufficiently general approximating functions any classification problem (any underlying probability distribution) can be solved with any desired accuracy, no practical method has been developed that guarantees to give acceptable solution in a general case.

1.3 Linear and quadratic discriminant functions

In certain cases, the optimal rule can be simplified to a more managable expression involving few parameters and a function which is relatively easy to evaluate. This is a very important consideration for practical application. We will show that this happens, for example, in the important case of normal (gaussian) distributions. To simplify the derivation we will consider the one-dimensional case: Let

$$p_i(x) = \frac{1}{\sqrt{2\pi}\,\sigma_i} l^{-\frac{1}{2}\left(\frac{x-\mu_{ji}}{\sigma_{ji}}\right)^2} \tag{13}$$

$i = 1, 2$. The Bayes rule then becomes

$$2 \ln \frac{p_1}{\sigma^1} - \left(\frac{x-\mu_1}{\sigma_1}\right)^2 \gtrless 2 \ln \frac{p^2}{\sigma_2} - \left(\frac{x-\mu_2}{\sigma_2}\right)^2 \tag{14}$$

and if $\sigma_1 = \sigma_2$ then

$$x(\mu_2 - \mu_1) \gtrless \sigma_1^2 \ln \frac{p_2}{p_1} \tag{15}$$

Similar expressions can be derived in the multivariate case. (σ_i is now replaced by the covariance matrix and μ_i by a vector.) In this special, but not uncommon, case computation becomes quite simple. In many practical situations, the conditions under which the rule has been derived do not hold, but the distribution is still reasonably close to normal or at least the resulting simple type of decision boundary separates the two classes sufficiently well. The application of linear or quadratic discriminant functions is then not fully justified and in general the resulting rule is not quite optimal but this may be outweighted by the simplicity of the rule. For this reason the linear and quadratic discriminant functions of the type described and their variants are the most commonly used classification rules (sometimes in a disguised form, such as use of tables to make decisions).

The major difference between the expression derived above for the gaussian case and the generally used methods is that in the gaussian case the meaning of parameters is related in a simple way to the statistical parameters of the considered distributions. In the more general situations, the parameters are simply constants which in a chosen sense optimize the performance of the rule. If we knew that the distribution is normal we would simply estimate the mean vectors and covariance matrices (from D_1 and D_2). In the general situation we search for the best values of parameters (using again D_1 and D_2).

The decision boundary generated by linear or quadratic discriminant functions (which is the solution of the equality case in the decision rule) is a linear or quadratic hypersurface. From this fact it is immediately obvious that the applicability of the method of linear or quadratic discriminant functions in the "pure" form is very limited; particularly multimodal situations (Fig. 12) cannot be satisfactorilly handled. For this reason, methods generalizing this approach and retaining its simplicity have been developed. One type of the most widely studied and used approaches are the piecewise-linear discriminant functions as described in the next section. Other methods are more directly related to the original gaussian problem, as for example the method of coalitions (Moore 1973) which assumes multimodal gaussian distributions in its derivation.

Linear and quadratic functions are the simplest of the polynomial functions. With increasing degree of the polynomial more and more complex surfaces (boundaries) can be described and in general any reasonable boundary can be, at least theoretically, approximated with any desired accuracy by one or more polynomials of sufficiently high degree. The cost for the increased effectiveness is the increased number of calculations that have to be performed for each classification and the complexity of evaluating the coefficients of the polynomial. These more general polynomial discriminant functions have been proposed for waveform processing for example by Specht (1967 a, b).

1.4 Piecewise-linear and piecewise-quadratic functions

When the underlying distributions are multimodal, the approach described in the previous section fails. The optimal decision boundaries in these cases can be approximated by a combination of linear or quadratic boundaries (Fig. 13). In fact these boundaries composed of linear or quadratic sections can be used to approximate any boundary with any desired accuracy if enough segments are constructed. The problem in general is, how to find these approximations. Since these methods are intended for use in complex situations we should be able to approximate boundaries even when we do not know beforehand how many modes there are; we would like to be able to construct the piecewise boundaries simply from the knowledge of the design set. This problem has not yet been solved completely. The existing methods in general sacrifice economic representation for efficiency of computation; for

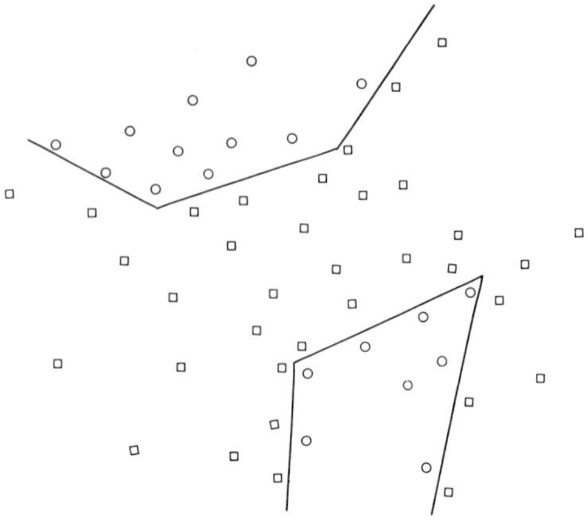

Fig. 13 Piecewise-linear boundaries. An algorithm using these boundaries for classification would achieve good separation of classes (providing the presented samples are representative of the overall distribution).

example boundaries are constructed only in specific directions for ease of calculation of parameters – such as parallel with axes, so that the real boundaries are not approximated economically (with a small number of parameters) (Fig. 14). For their simplicity of implementation and generality these methods are the subject of much research.

1.5 Nearest neighbor rules

Let y be a sample with unknown classification, x its nearest neighbor from $D_1 \cup D_2$. It is natural to assign y to that class to which x belongs. This rule is known as the nearest neighbor (NN) rule. In a more sophisticated form we find k (instead of only one) nearest neighbors and assign x to that class to which most of these samples belong – the k-NN rule (Fig. 15). This conceptually very simple rule has been shown to be surprisingly effective; for D large enough the performance is not worse than twice the optimal one:

$$\min_{X_1,X_2} E(X_1,X_2) \leq E(R) \leq 2 \min_{X_1,X_2} E(X_1,X_2) \quad (16)$$

(the expected relative number of errors is not worse than twice the minimum, obtainable with the optimal rule). It has recently been shown (Tomek 1974 a, 1976 b) that very simple preprocessing (consisting of a repetitive application of the k-NN rule to D and elimination

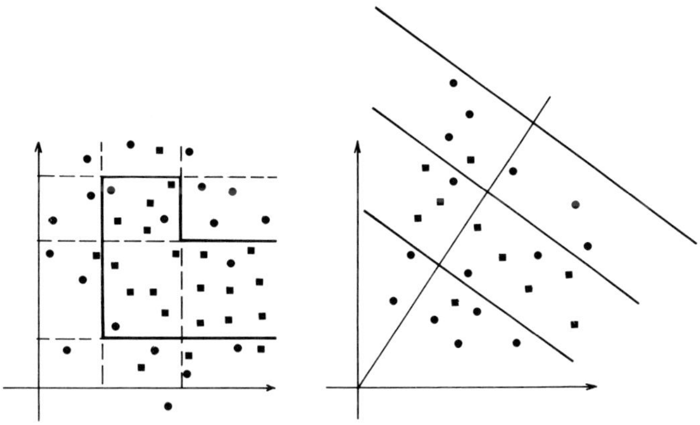

Fig. 14 Piecewise-linear boundaries of restricted types (from Meisel [16]): a) boundaries alligned with axes, b) boundaries parallel with a general direction.

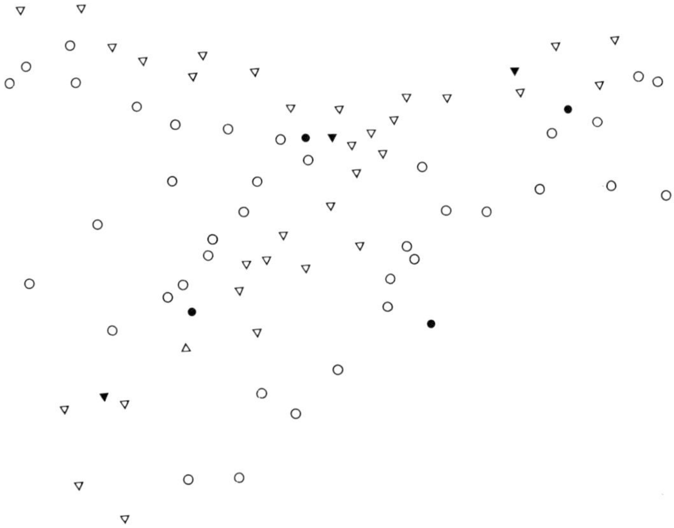

Fig. 15 Classifying a sample by its 3 nearest neighbors. Full symbols are examples of classification obtained by this rule. Note how difficult it would be to generate the decision boundary by other means.

of the incorrectly classified samples) could result in the generation of a new (reduced) design set which classifies better than the original design set (Fig. 16).

The advantage of *k-NN* rules is that they are known to guarantee good classification and are quite general. We have mentioned that methods based on approximations of density functions theoretically guarantee optimal performance in the limit, and so in fact do piecewise-linear methods. The major difference is that the proof of this property is not constructive in the case of these methods and no algorithm, which in a general case guarantees good performance, is known. In the case of the *k-NN* rules this is not so since the method itself is constructive and does not require any intervention of the user; no subjectively chosen parameters, except for the value of *k*, have to be supplied.

The main disadvantage of *k-NN* rules is their computational inefficiency. All samples from the design set have to be stored and to classify a new sample its distance to all the samples in the design set has to be calculated. This is unmanagable for larger design sets particularly in higher dimensional spaces. Because of the attractiveness of *k-NN* rules,

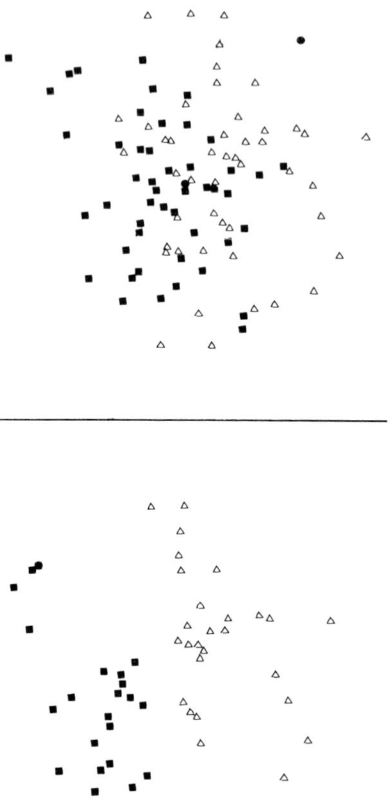

Fig. 16 An example of a design set before and after processing to improve nearest neighbor classification (from Tomek [30]) Note how the originally overlapping distributions are separated by the process of elimination.

these problems have been the subject of much research and at present powerful methods are known which reduce the above drawbacks.

The following two directions have been used to develope methods which improve the efficiency of *k-NN* rules. The first direction entails reduction of the size of the design set so that its representativeness is not affected (in general the idea is to eliminate all samples which do not participate in the generation of the decision boundary). This reduces both the storage requirements and the number of calculations that have to be performed (Hart 1968). The second method is to store the resulting design set in a way which permits the location of the nearest neighbor in a small number of calculations so that not all distances have to be calculated (Fukunaga, Narendra (1974)).

Although currently not very popular in practical applications, the *k-NN* rule can be expected to become more widely used in the more difficult situations as soon as the recent results become known to the general public. It can also be expected that the increased interest in the method will result in the discovery of even more powerful processing methods.

1.6 A comparison of the described methods

Of the five methods described above, three seem to be particularly attractive. The linear and quadratic discriminant functions are very simple and should be used wherever their use guarantees acceptable results. The piecewise-linear and piecewise-quadratic methods are potentially both effective and efficient. More research is needed to develop generally applicable algorithms with guaranteed efficiency and effectiveness in unrestricted situations, in which the simple discriminant functions cannot be used. Finally the *k-NN* methods will probably become more popular in the general situations in which they guarantee good classification. More research is still needed to make these methods computationally more efficient. Approximation of density functions is the best alternative perhaps only in those cases where the quantitative knowledge of probabilities is required.

2. Classification with Reject Option

In many applications, a decision rule of one of the types described above is not good enough. The reason for this is that often we will accept a classification of a sample into a class only when we are reasonably sure that it is correct. In our statistical setting this can be rephrased as follows: A classification is accepted only if the likelihood ratio (defined below) exceeds a prespecified threshold. Samples which cannot be classified with this degree of certainty (likelihood ratio does not attain the threshold value) are not classified at all – they are rejected. The likelihood ratio $l(x)$ is defined as

$$l(x) = \frac{p_2(x)}{p_1(x)}$$

and the reject option classification can in the case of the Bayes rule be formulated as follows:

$$\text{C1} \qquad (-\infty, t_1) \qquad (17)$$

assign x to R if $l(x) \in (t_1, t_2)$

$$\text{C2} \qquad (t_2, \infty) \qquad (18)$$

Here $t_2 > t_1$ are thresholds, R the reject class (Fig. 17). (For a more detailed treatment of the reject option, see Chow (1970).

The implementation of this rule, very useful and natural in biomedical applications, is quite simple in particular if approximative methods from section b are used. It can also be efficiently implemented when the k-NN rule is used (Tomek 1976 a). This can be done by finding samples which belong to the reject class (Fig. 18). The method is based on the fact that the volume containing k samples from a given class can be used to approximate the probability density in the given region.

III. Examples of Approaches to ECG and PCG Evaluation

A. Introduction

Although much research has been devoted to automatic evaluation of various types of biomedical waveforms and many papers published on programs written for this purpose, only a few of the ECG evaluation programs have found practical application. There are many reasons for this strange situation. Most of the techniques recording phenomena associated with cardiac events do not have established standards (ECG is the exception) and in fact computer processing is an important stimulus for the development of such quantita-

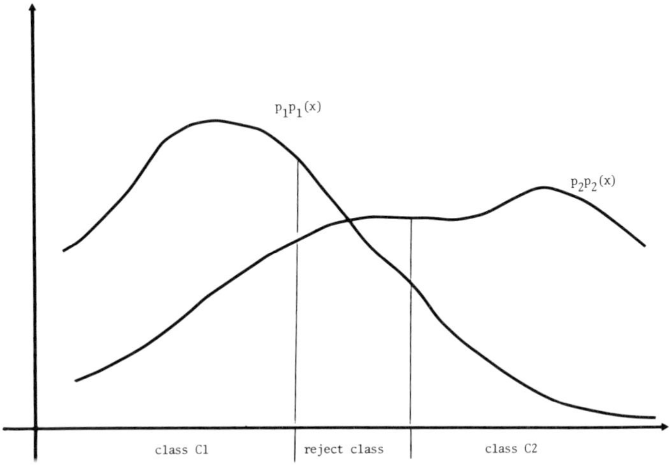

Fig. 17 Classification with a reject option: Samples close to the decision boundary are excluded from classification since they cannot be assigned with sufficient certainty

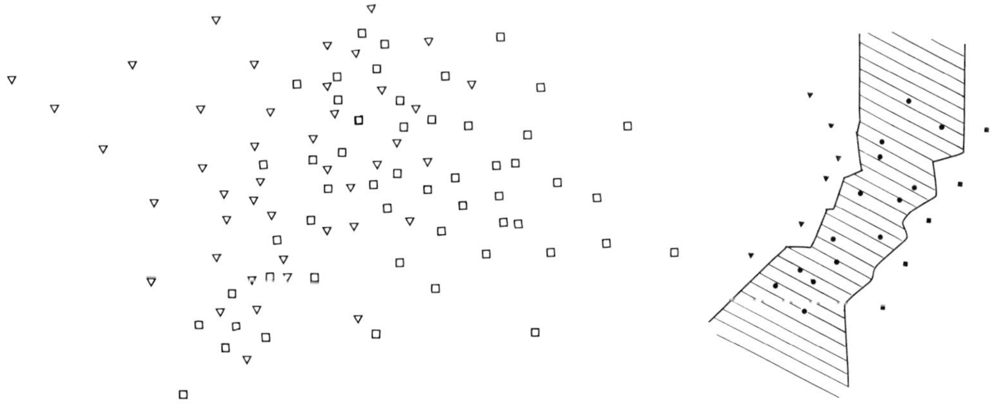

Fig. 18 Original design set (part a) and the generated partition into two classes and a reject class (b) for nearest neighbor reject classification. The boundary generated by the boundary samples is sketched and the reject region shaded. Samples falling into this region are rejected from classification (they are left unassigned).

tive criteria; there is also a certain lack of confidence of the medical profession (even to well tested ECG programs known to perform well) partially due to insufficient understanding.

Then there is the cost of practical implementation which includes many changes in existing methods of recording, storage and handling of records, as well as the development of programs working reliably in real life situations; many techniques are used relatively rarely (for example the BCG) and there is no sufficient reason for the introduction of expensive computers into the process. This could, however, change with the introduction of automatic evaluation since a number of methods are not used simply because there is not enough personnel experienced in their quantitative use. It seems that the lack of coordination of research is also a factor in the random character of development.

We have mentioned, in our disscussion of 'classification methods' that there are a number of criteria that can be used to derive decision rules. We have also introduced one of them – the minimum expected risk (cost) criterion. We will now show how this criterion can be used in the classification of medical waveforms.

Most of the graphic cardiac techniques are (or could be) used in all the following situations: a) screening of large populations, b) diagnosis of suspected abnormal cases, c) monitoring of seriously ill patients. In each of these situations the required criteria and the corresponding approaches are different.

In screening, we check large populations of mostly normal subjects and want to identify all possibly abnormal cases. The suspected abnormal subjects are then examined in more detail. When we make an error and classify a normal subject as abnormal, the error is not as serious as if we classified an abnormal subject as normal and thus eliminated him from further examination. Our criteria should thus be biased in this direction; in other words we will use a different set of costs. Note also that in screening our algorithm only has to be able to distinguish two classes – normal and abnormal – more sophisticated algorithms would be wasteful.

In diagnostic evaluation, we want to make an accurate diagnosis, using a number of classes (normal and a number of abnormal conditions). Our cost constants will be different.

In monitoring of seriously ill patients we know the condition but want to catch certain dangerous states and the situation is again different. In fact, even the signal is different; since the patient is monitored continuously the signal is often very noisy and we have to make sure that our algorithm is unsensitive to noise.

Note that the use of the minimum expected cost criterion allows us to design consistent rules; all boundaries are shifted when cost constants are changed, but they are shifted appropriately. If we attempted to change parameters (i.e. move the decision boundaries) used by the algorithm subjectively, we would not have a guarantee of consistency. This shows the advantage of using a systematic approach.

We are going to describe several examples of programs for computer processing. Some of them are more or less generally accepted and have been carefully tested and their performance compared against the performance of cardiologists (and found in most respects at least as good as theirs). Others are only proposals never practically implemented; it may be, however, that some of them will eventually be accepted when the gap of understanding between the medical and technical communities is bridged.

B. ECG Evaluation

ECG evaluation by algorithmic methods is the most developed of all graphic cardiological methods. The reasons for this are the widespread use of the ECG (and the consequent high demand) and the existence of a large number of generally accepted criteria; although there are differences between individual cardiologists in the interpretation of individual records the majority of situations is handled in the same way. These two reasons are interrelated since frequent use leads to the discovery and acceptance of standards and this in turn increases the attractiveness of the method. There is a limit on how many ECGs can be evaluated; it is given by the number of people qualified to do it. This limit would be quite different if computers were accepted as reliable in at least certain situations. The ECG could then be used even more widely. This and the necessary detailed analysis which accompanies the development of each such system would then lead to an even better understanding and more objective interpretation of the ECG.

We will shortly describe several approaches to automatic evaluation of the ECG, starting with two which have found practical application and continuing with another which, al-

Table 1 Screening Results of Selected Heart- Stattion Evaluations

Program	Site	Date	Leads	Correct Records	Normal Records	Correct %	Abnormal Records	Correct %
USPHS	George Washington University Hospital	1966	ECG	73	148	72	598	99.2
USPHS	Hartford Hospital	1967	ECG	87	113	100	253	81
USPHS	Hartford Hospital	1967	ECG	78	276	80	52	67
USPHS	George Washington University Hospital	1968	ECG	81	178	97.3	469	96.6
USPHS and Queen's	Queen's University Teaching Hospital	1969	ECG	82	112	86	88	
USPHS and Queen's	Hotel Dieu Hosp. and Kingston General H.	1970	ECG	85	762	90	1257	96
Mt. Sinal	Mt. Sinal Hospital	1967	ECG	91	966	91	1064	94
Mayo Clinic	U. of Washington and Group Health Hosp.	1970	VCG	74	289	75	84	76
Latter-Day Saints H.	Latter-Day Saints Hospital	1969	VCG	95	195	99	92	91
Osaka Center	Osaka Center for Adult Diseases	1966	VCG	82	100	80	105	84
Iowa Methodist	Iowa Methodist Hospital	1967	I, II, V2		1708	99.7	3090	91
Mt. Sinal and Mayo	l'Hospital Bruaman	1970	ECG	85	432	91.5	1303	
			VCG	81	105	83		
		1971	VCG	67	2589	78	603	95.6
			ECG	89	1487	95	5334	
Mt. Sinal and Mayo	CHU Bretonneau a Tours	1971	ECG	85	73		427	
			ECG + VCG	87	73		427	
Royal Infirmary	Royal Infirmary Glasgow	1971	VCG	81	205	73	419	85
			ECG	82	205	73	419	86
VA Research	VA Hospitals	1971	VCG	74	597	90	2005	89

though potentially more powerfull, has not yet been accepted to the same degree. For a survey of practical use of different ECG programs, see Tables 1 and 2. The reader interested in a more detailed review can consult, for example, the paper by Cox et al (1972) which contains a very informative section on ECG programs.

1. The USPHS and the IBM-Mayo Clinic Programs

These programs (Weihrer 1970, Alexander 1970) are the two that gained the largest acceptance in practical use. They are designed similarly and we are going to describe the USPHS (US Public Health Service) program developed by Dr. Caceres and his group only.

The program uses the 12 lead ECG. It is not taken by a standard ECG equipment but one designed specifically to be interfaced with a computer. The main distinguishing features are:

(i) its data acquisition unit is capable of generating a code to identify the ECG and the patient, and

(ii) it is equipped with a timer which automatically provides records of a standard (5 seconds) length. The ECG is transmitted to the computer through a telephone and also recorded on magnetic tape and in a visible form. The signal is sampled at the rate of 500 samples/second and converted from the analog into the digital form.

The USPHS program consists of two parts (as most ECG programs do); in the first part the important features are recognized, measured and outputed and in the second part a diagnosis produced. A typical output is shown in Fig. 19. In the second part, the tabular method using logical functions is used to classify the record on the basis of characteristic values found in part 1 and a table of thresholds stored in tabular form. These latter constants are stored in computer's memory and can be modified by the user.

A comparison of the performance of a cardiologist and the program is shown in Table 3. It should be kept in mind that criteria used in

Table 2 Computerized processing of electrocardiograms in Canada (1972)

Data acquisition							
Site	Program	Input units	Lead system	Record lenath sec	Sampling rate/sec	Computer system	Volume (monthly)
Halifax Dalhousle University	ECAN/D (soalar lead) Dalhousle (vector lead)	Hawlett-Packard Marquette	12 lead Frank	4 15	500	Xerox Sigma 5	3000
Quebec Institute of Cardiology	USPHS/D	Cambridge	12 lead	3.72	500	CDC 1700	1050
Montreal Notre-Dame Hospital	Mayo	Marquatte	12 lead Frank	2.5 10	1000	IBM 1800 with MPX	
Montreal University of Montreal	Dalhousle	Hewlett-Packard	12 lead Frank	15 90	500	DEC PDP9 CDC 6600	120
Kingston Queenis University	USPHS (QV) and Dalhousle	Marquette	12 lead Frank	5 10	500	IBM 360/50 or Burrouahs 6700	700
Toronte Hospital for Slok Children	Mayo (68)	Electronlos for Medcine DR-8	Frank	10	1000	DEC PDP8 and IRM 360/40	
London Victoria Hospital	Mayo (68)	Marquette	12 lead Frank	 10	 250	IBM 360/50 with MVT	120
Winnipeg University of Manitoba	USPHS/D	Marquette	12 lead	2.3	500	CDC 1700	
Saskatoon University of Saskatohewan	USPHS (QV) and Mayo (68)						
Edmonton University of Alberta	USPHS/D Mayo (68)	Hewlett-Packard	12 lead Frank	2.3 10	500 250	Hewlett-Packard 2100A	
Vancouver General Hospital	Vancouver	Tapa carts	Frank			Dual IBM 360/67 with MTS	

Table 3 Physician's reasons for disagreement with computer in 157 abnormal electrocardiograms

Reason	Number
(a) Inadequate ECG diagnostic criteria used by computer	51
(b) Quantitative disagreement with computer measurements	53
(c) Disagreement with semantics of computer diagnosis	37
(d) Artifact	22

Table 4 Computer classification compared with cardiologist classification

	Abnormal (n = 598)	Normal (N = 146)
Complete agreement	436 (73%)	107 (72%)
Agreement of abnormality but disagreement on nature of the abnormality	157 (26%)	—
Disagreement	5 (0.8%)	41 (28%)

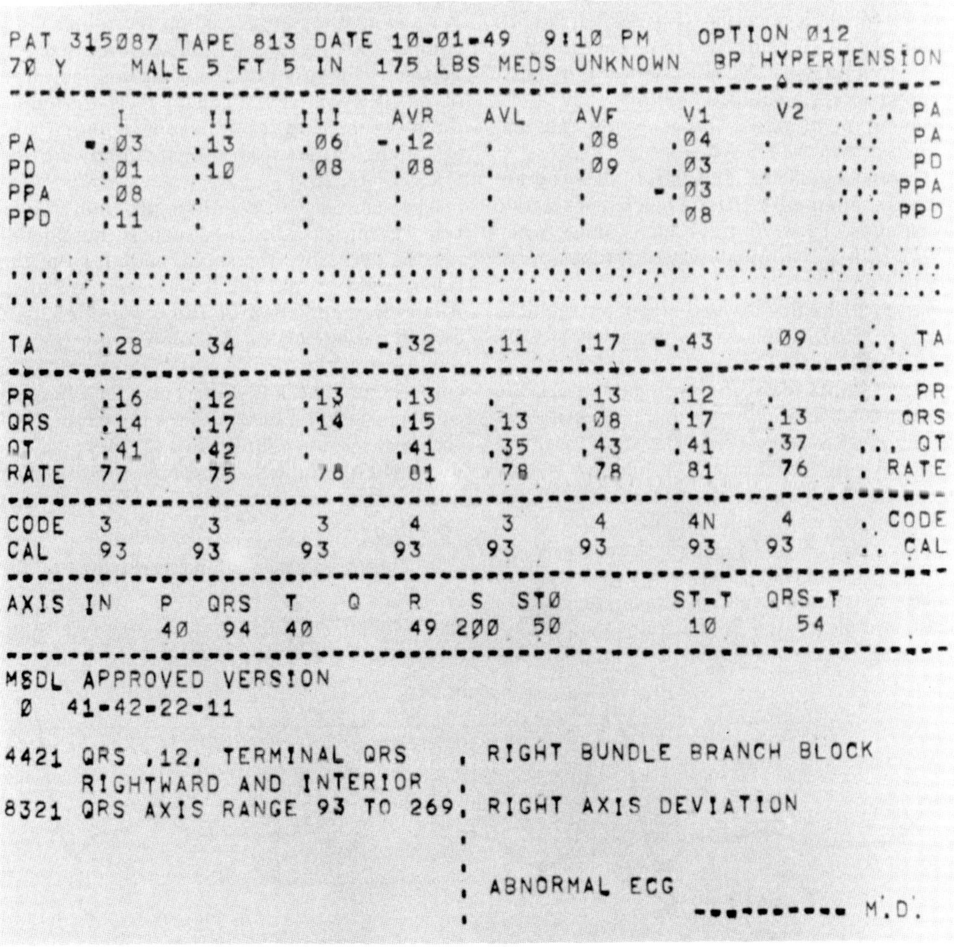

```
PAT 315Ø87 TAPE 813 DATE 1Ø-Ø1-49  9:1Ø PM   OPTION Ø12
7Ø Y    MALE 5 FT 5 IN  175 LBS MEDS UNKNOWN  BP HYPERTENSION
------------------------------------------------------------
        I      II     III    AVR    AVL    AVF    V1     V2     .. PA
PA    -.Ø3    .13    .Ø6   -.12    .      .Ø8    .Ø4    .     ... PA
PD     .Ø1    .1Ø    .Ø8    .Ø8    .      .Ø9    .Ø3    .     ... PD
PPA    .Ø8    .      .      .      .      .     -.Ø3    .     ... PPA
PPD    .11    .      .      .      .      .      .Ø8    .     ... PPD

............................................................
............................................................

TA     .28    .34    .     -.32    .11    .17   -.43    .Ø9   ... TA
------------------------------------------------------------
PR     .16    .12    .13    .13    .      .13    .12    .     ... PR
QRS    .14    .17    .14    .15    .13    .Ø8    .17    .13       QRS
QT     .41    .42    .      .41    .35    .43    .41    .37   ... QT
RATE   77     75     78     81     78     78     81     76    . RATE
------------------------------------------------------------
CODE   3      3      3      4      3      4      4N     4     . CODE
CAL    93     93     93     93     93     93     93     93    .. CAL
------------------------------------------------------------
AXIS IN   P  QRS  T    Q    R    S  STØ        ST-T  QRS-T
          4Ø  94  4Ø        49  2ØØ  5Ø         1Ø     54
------------------------------------------------------------
MSDL APPROVED VERSION
 Ø  41-42-22-11

4421 QRS .12, TERMINAL QRS    , RIGHT BUNDLE BRANCH BLOCK
     RIGHTWARD AND INTERIOR   .
8321 QRS AXIS RANGE 93 TO 269, RIGHT AXIS DEVIATION
                              .
                              .
                              . ABNORMAL ECG
                              .            ---------- M.D.
```

Fig. 19 A typical printout generated by the USPHS ECG program (from Weihrer et al [36])

this case are such that as few abnormal cases as possible are missed, classifying normal cases as abnormal is thus preferable to classifying abnormal records as normal. Table 4 summarizes reasons for disagreement in the abnormal ECGs. This table is a typical tool providing feedback via the comparison of desired and actual results. It makes it possible to introduce changes of criteria in the program and thus gradual elimination of differences between the cardiologist and the computer program. It is anticipated that the program will be extended to the processing of such data as heart scunds, blood pressure curves etc.

We are now going to describe in more detail the two main parts of the program:

a. The "recognition" part of the program

has three main sections (Fig. 20 shows an idealized ECG for illustration of notation):

1) rhythm analysis and selection of the "best" complex for a detailed analysis,
2) definition of waves (onsets, ends and amplitudes) in the chosen complex,
3) search for artifacts. This part of the program uses a number of criteria (such as excessive baseline deviation, noise etc.), which, when not satisfied terminate further processing of the signal.

The record is first divided into shorter blocks (0.37 s long) which are searched for QRS complexes. To find these complexes, the most negative slopes (derivatives) in each block are found as well as amplitudes and ranges of these possible QRS complexes. The most negative derivative in the currently processed lead is then used as a basis for determining all the QRS complexes present (for this purpose, all the previously found negative slopes are compared with a threshold obtained as a fraction of the maximum negative slope). In each of these complexes, the R peaks are then found. The lengths of consecutive R-R intervals are compared to check for arrhytmia. The deviation of the baseline between consecutive R-R's is then examined and the "best" complex (one which has the least deviation) determined on the basis of this information.

The best complex is used to determine the exact location of the Q, R and S points providing its deviation from the baseline is smaller than a threshold value. If this condition is not satisfied, the lead is considered technically unsatisfactory. The logic of QRS detection is much too complex to be described in this chapter. To illustrate this, let us mention that a description of the identification of the P wave requires 3 pages in the paper from which this summary has been compiled.

After the completion of QRS identification, the program searches for the occurrence of a P wave in a region preceding the found QRS complex. The endpoints of this region are determined by the heart rate and the configuration of the QRS complex. When the P wave is found, the program finds the T wave in a similar way (searching in a specific region following the QRS complex).

Throughout the recognition program a number of empirical constants, such as those defining the derivative threshold, endpoints of the region used to find the P wave etc., are used. After the completion of the recognition part certain parameters of the complex are printed out, to allow the cardiologist to reconstruct the complex and check the interpretation obtained by the program if he so desires. An example of such a printout is shown in Fig. 19. A code is used to output other technical information such as excessive noise, excessive scale (amplitude of signal), etc.

b. For the purpose of interpretation of the record

the original data are now condensed; only those values essential for diagnosis are re-

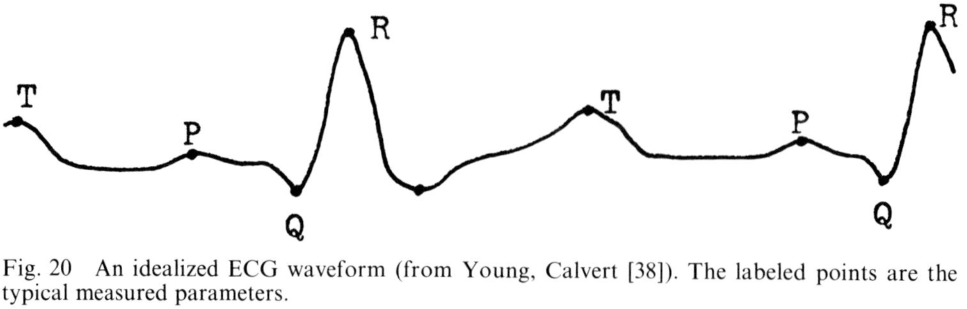

Fig. 20 An idealized ECG waveform (from Young, Calvert [38]). The labeled points are the typical measured parameters.

tained. They are the parameters to be used in the diagnostic process in the next part of the program. The condensed data are then processed by a decision tree – tabular method. This means that the identified feature vector is compared with values stored in "work sheets". Processing is done essentially by application of logical AND OR functions which direct the way the search tree is examined. Let us note, at this point, that in terms of our terminology used to describe classification methods this approach corresponds to a special type of piecewise-linear boundaries shown for a simplified two dimensional case in Fig. 21. This illustration shows how limited configurations of boundaries can be obtained by this method. The use of this approach has been dictated by the authors' attempt to imitate classification as performed by cardiologist (according to what cardiologists consider to be an adequate description of the process).

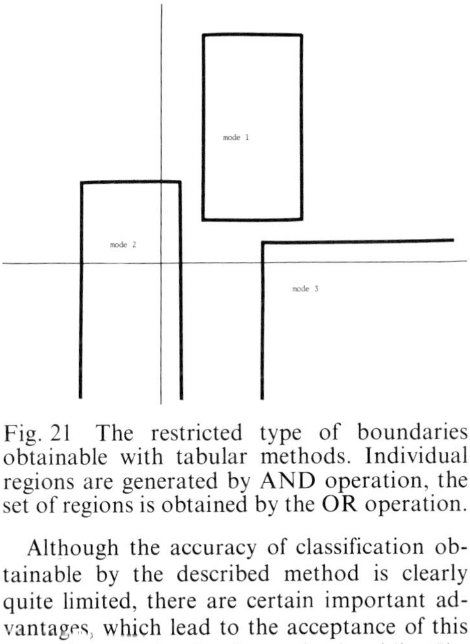

Fig. 21 The restricted type of boundaries obtainable with tabular methods. Individual regions are generated by AND operation, the set of regions is obtained by the OR operation.

Although the accuracy of classification obtainable by the described method is clearly quite limited, there are certain important advantages, which lead to the acceptance of this type of processing by the medical public. The program essentially consists of a fixed part designed to find all important features of an ECG record and another part which makes decisions (diagnosis) on the basis of the data found in the first part. The second (diagnostic) part of the program works by making decisions using a decision tree and uses internally stored constants. This part can be modified, the user having the option to replace standard criteria (the structure of the decision tree and the corresponding constants) by his own.

In this way the program can also be updated to reflect the most recent results of cardiological research. Regional differences and individual preferences can be incorporated into the program by modification of work sheets. The cardiologist also understands what the program is doing since the program is following the same routine as that used by the doctor, who can thus check the result using the same method. There is also a psychological factor, namely that the process is not based on data forms and methods too abstract and without immediate physical meaning for a medically trained person.

When interpretation is completed, the conclusion reached is printed out.

c. A separate program makes possible computer plotting of the recorded signal

for the use by the cardiologist (Fig. 22). Its purpose is to save time (an experienced technician needs 4 to 5 minutes to prepare a record while the plotter needs 1 minute and 20 seconds) and to eliminate the necessity to use specially trained personnel (to eliminate mix-ups and improper selection of ECG complexes). The plot also presents the record in the form in which it is processed by the program – that is discrete. This, however, has not been appreciated by doctors who are used to smooth analog records.

Both programs mentioned in this section have performance comparable to that of cardiologists in most situations. There remain, however, certain problems which have not been solved satisfactorilly, particularly arrhytmias. This defect can be explained in part by the difficulty of identification of the P wave in a number of records. This is a common problem of all ECG programs. It is the subject of intensive research.

We are now going to describe another program which uses a different approach – more along the scheme outlined in our earlier presentation of classification rules.

2. The US Veterans Administration Research

The group of Dr. Pipberger used a large data base of more than 40000 records from eight hospitals of the US Veterans Administration in their extensive research into computerized ECG diagnosis. Their attention has focused on

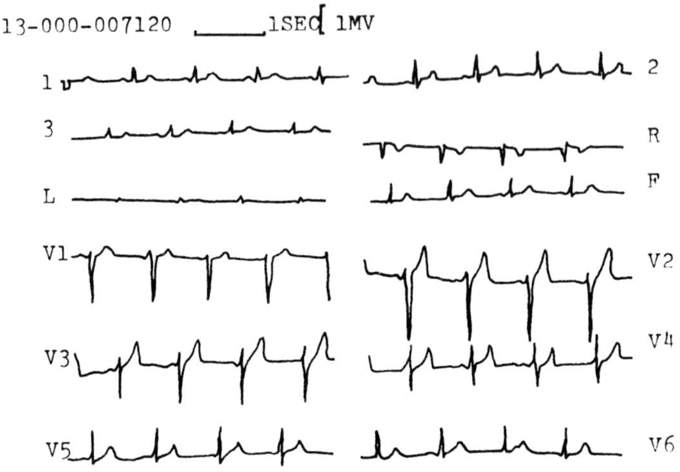

Fig. 22 An example of a digitally plotted ECG as provided by the USPHS program (from Weihrer et al [36])

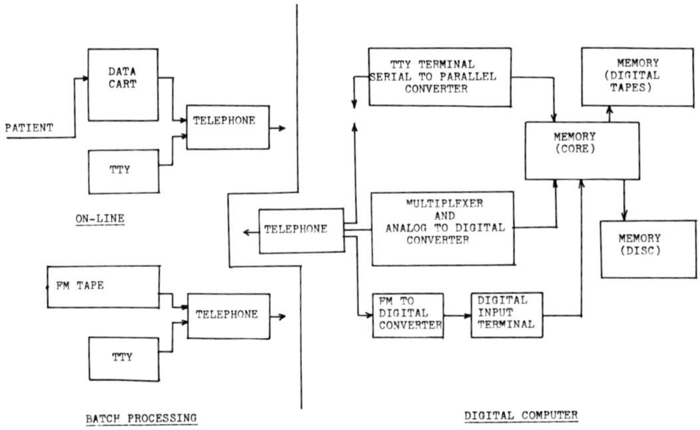

Fig. 23 Data flow in computerized ECG diagnosis (from Klingeman, Pipberger [13])

the use of more powerful methods of classification, particularly those based on the assumption of normal distribution of samples, described shortly above. In this respect their approach differs substantially from the program described in the previous section.

The organization of data acquisition and processing is shown in Fig. 23, and is discussed in more detail in Klingeman, Pipberger (1970). In our presentation of their work, we are not going to deal with the organization of the program or a more detailed description of their use of normal distribution assumption. We will, instead, deal with another important aspect of their research, namely the way in which features (components of sample vectors) were chosen to guarantee that as few quantities as possible have to be used to achieve a satisfactory performance. In other words how features are selected in order to achieve effective and efficient classification.

In the beginning phase of the research, the group decided to find out which of the large number of features recommended by various authors are most efficient discriminators and should be used by their program. A factor in this choice is also the amount of calculations required to obtain these features. Some features are obtained by simple subtraction – such as amplitudes and durations; other features require non-trivial calculations – such as eigenvalues. The calculation of these features would slow down processing significantly and this is an important consideration.

The program uses the Franke ECG system. It was decided to compare more than 300 features. The work started by applying the linear discriminant function method to the ECG record, digitized at evenly spaced points from the QRS and ST-T complexes. These vectors were compared (presumably by calculating the euclidean distance) against mean vectors representing various abnormalities. After some experimentation, it was concluded that this approach is not satisfactory and that the reason for this is that the number of samples in the design set is insufficient to determine reliable mean vectors and stable decision boundaries (note that the larger the dimension of the sample space the larger the number of samples necessary for a truly representative design set). It has been estimated (Pipberger 1970) that the number of samples should be of the order of the square of the number of features used. The maximum number of components that can be used for classification with any hope of success is thus limited by the size of the design set. With a relatively small design set, a large number of features does not guarantee a good performance.

On the other hand, if we know how many components the pattern vector should have (this may in some cases be given by our knowledge of the physical meaning of the record), we have to collect at least a certain minimum number of classified patterns in order to obtain a representative design set). For this reason, all other features recommended in the literature were evaluated for each record. These features were then examined by comparison of the ranges of their values in the different classes. Their number was then reduced to the 30 most promising ones. These were then compared in more detail and the most efficient ones (as judged by discriminant function analysis) retained for use in the program. Mucciardi and Gose (1970) examined this problem separately and concluded that about 15 features guarantee a performance which cannot be significantly improved by adding new features.

It has been mentioned above that the diagnostic program itself uses a statistical approach to classify the ECG; it is not simply an imitation of the cardiologist. This approach has not, however, achieved the acceptance gained by the two programs mentioned earlier, partly exactly because of this fact. There have been other attempts to use statistical theory to classify ECGs, some of them using more sophisticated approaches than linear discriminant functions. So Specht (1967 b) proposed polynomial discriminant functions, Mucciardi and Gose (1970) multimodal normal distributions, Cady et al (1961) the use of Fourier expansion etc. These methods have not achieved practical acceptance either, undoubtedly partly because of the impossibility to express their decision processes in terms compatible with ECG interpretation as performed by cardiologists. It can be expected, however, that the theoretically more powerful methods will eventually be accepted when the "cultural barrier" between the medical profession and the technical community accustomed to statistical terminology and methods is overcome. A comprehensible presentation of the underlying principles and demonstration of the superiority of these methods will speed up this process.

C. Processing of Phonocardiograms

We are going to describe two methods for the processing of phonocardiograms (PCG) in order to show how different approaches can be used to attack the same problem.

To choose a method for the recognition part of the PCG processing program is not such a simple decision as in the case of the ECG. The reason is that while it is generally accepted that the P, Q, R, S and T points (which are more or less clearly defined on the original record) contain practically all information about the ECG, in the case of the PCG the information is not in specific points but rather in specific local frequency components (Fig. 24). To extract this information, we have to be able to describe their location quantitatively and this can be achieved in a number of ways.

1. Energy Content Approach

Perry et al (1967) described a program which uses the energy contained in the signal to extract the most important portions and interpret the record. They use a microphone with a flat frequency response in the range (50–1500 cps) to measure the PCG and then remove the unnecessary components by a bandpass filter 100–500 cps. A/D conversion is performed by sampling at the rate of 3000 samples/second. The ECG signal which is used by the recognition part of the program is recorded simul-

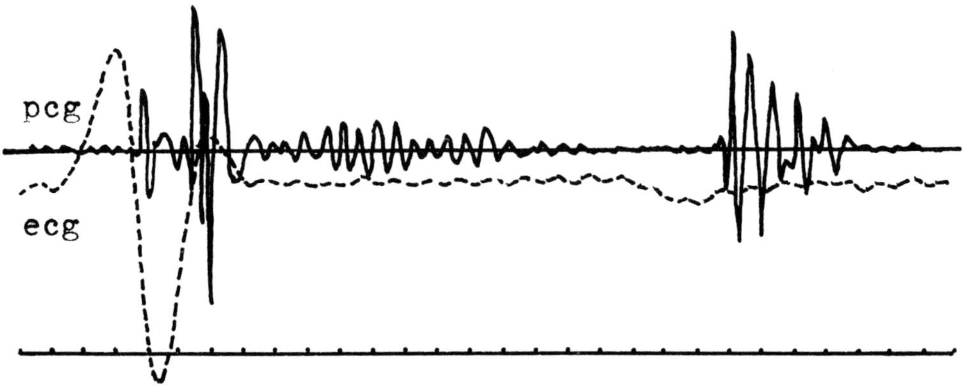

Fig. 24 A phonocardiogram and a reference electrocardiogram showing the relative position of their important sections (from Perry et al [20])

taneously (Fig. 24). Signal data of 2.4 seconds is stored in digital form. The recognition program uses the R wave of the ECG as a reference point; it is used to calculate regions in which a search for individual heart sounds is performed.

The whole cycle is divided into segments and recorded vibrations described by the following parameters: the time of onset, duration, frequency and amplitude (relative with respect to the amplitude of the second sound). From each chest location, three heart cycles are analyzed. The description obtained by the recognition part and the interpretation are printed out (Fig. 25) and stored on magnetic tape. Diagnosis is made by a tabular method imitating the process used by physicians. It uses amplitudes, durations and onsets of sounds for their classification as murmurs, ejection sounds, 3rd and 4th heart sounds etc., to arrive at one out of a total of 19 diagnoses.

2. Frequency Components Approach

Litwin and Begon (1974) describe an approach which differs from the one described above. They record the PCG plus an ECG and the carotid pulse or an apexcardiogram. The PCG is again sampled at 3000 samples/second. It is divided into blocks ("articles") 1.2 sec long. The length is chosen to correspond approximately to that of an average heart cycle. The beginning point of each article is the beginning of the QRS complex of the ECG. Each article is divided into 26 successive segments of equal length (128 points) and for each segment the spectrum is calculated using the Fast Fourier Transform (which requires the total number of points to be a power of 2: $N = 2^n$; in our case $n = 7$).

Segments are characterized by two parameters which can be used to approximately reconstruct the original signal: the maximum amplitude of the power spectrum of the segment and the corresponding frequency (Fig. 26). These parameters are then used as the components of a feature vector. Since each article consists of 26 segments and each of them is characterized by two numbers, the approach uses 52 dimensional feature vectors. Litwin and Begon (1974) describe an example of typological analysis of these vectors by factorial analysis, using a set of 67 observations (phonocardiograms). Results of this example show that it is possible to distinguish between the two analyzed types of diseases. These preliminary findings can be used as a part of a basis for a program for automatic classification of phonocardiograms. Clearly much work remains to be done to develop a program of general applicability and interest.

D. Concluding Remarks

We are not going to describe any more waveform processing programs, although a number of such programs exist for most of the used types of waveforms (for example blood pressure contours and ballistocardiograms). These programs are based on similar principles and most of them are not developed to a point where their practical use could be contemplated. From

INSTRUMENTATION FIELD STATION--HEART DISEASE CONTROL PROGRAM
COMPUTER PROCESSED PHONOCARDIOGRAMS

PATIENT-0507 REEL-0008 SEX-M AGE GROUP-08 HEIGHT GROUP-05
CHEST CIRCUMFERENCE GROUP-08 WEIGHT GROUP-06
PROCESSING DATE 06-18-65
MICROPHONE LOCATION

COMPLEX	1	2	3	1	2	3	1	2	3	1	2	3
S1A ONSET	12	-	1	1	0	2	0	0	19	14	0	0
AMP	10%	-	1%	20%	1%	1%	10%	13%	1%	46%	50%	1%
S1B ONSET	41	57	28	-	-	-	-	41	-	-	48	-
AMP	10%	1%	1%	-	-	-	-	17%	-	-	1%	-
S2A ONSET	271	271	268	268	269	271	267	266	261	270	265	274
AMP	100%	100%	100%	100%	100%	100%	100%	100%	100%	100%	100%	100%
S2B ONSET	296	306	312	308	300	312	-	287	-	300	284	308
AMP	119%	39%	22%	33%	38%	45%	-	42%	-	13%	13%	1%
S3 ONSET	-	-	-	-	-	-	-	-	-	-	-	-
AMP	-	-	-	-	-	-	-	-	-	-	-	-
S4 ONSET	-	-	-	-	-	-	-	-	-	-	-	-
AMP	-	-	-	-	-	-	-	-	-	-	-	-
RATE	96	68	96	96	97	97	107	107	79	103	102	102
SM1 ONSET	-	-	-	-	-	-	91	-	-	-	-	-
DURATION	-	-	-	-	-	-	40	-	-	-	-	-
% SYSTOLE	-	-	-	-	-	-	12%	-	-	-	-	-
CPS MEAN	-	-	-	-	-	-	125	-	-	-	-	-
AMP ONSET	-	-	-	-	-	-	19%	-	-	-	-	-
AMP MIDDLE	-	-	-	-	-	-	9%	-	-	-	-	-
AMP END	-	-	-	-	-	-	5%	-	-	-	-	-
DM1 ONSET	-	404	-	-	-	-	-	-	-	-	-	-
DURATION	-	20	-	-	-	-	-	-	-	-	-	-
% DIASTOLE	-	10%	-	-	-	-	-	-	-	-	-	-
MEAN CPS	-	350	-	-	-	-	-	-	-	-	-	-
AMP ONSET	-	27%	-	-	-	-	-	-	-	-	-	-
AMP MIDDLE	-	28%	-	-	-	-	-	-	-	-	-	-
AMP END	-	27%	-	-	-	-	-	-	-	-	-	-

INTERPRETATION: NORMAL PCG

R. O. ARTIFACT. DIASTOLE, 2RICS, 2ND COMPLEX
R. O. ARTIFACT. SYSTOLE, 4LICS, 1ST COMPLEX

Fig. 25 A printout for a computer processed phonocardiogram (Perry et al. [20])

the examples, we have shown it can be seen that a particular problem may be approached in many different ways – both the recognition of features and interpretation. Differences in the recognition part often lead to different representations of the record, some of which may be better suited for interpretation and/or recognition. Differences in classification methods may also result in better or worse performance.

It has been suggested why the theoretically less powerful methods are more popular than the more sophisticated ones. It is possible that the unwillingness to give a chance to the theoretically better methods is the underlying reason for some objections to automatic processing. The less powerful methods have a limited possibility of approximating optimal classification.

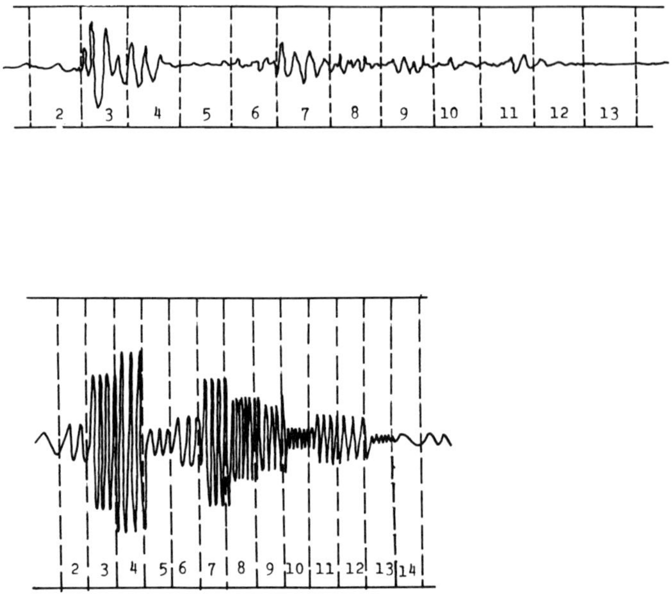

Fig. 26 A phonocardiogram and its approximation obtained from the maximum amplitude component (from Litwin, Begon [14])

Cardiologists generally assume that they interpret records more or less on the basis of certain accepted criteria (such as the P, Q, R, S and T points in the ECG) while in fact this may not be quite so. When their interpretative routine is supposedly translated into a program, it does not achieve exactly their performance. Some cardiologists then claim that the process is in fact intutitive and cannot be translated into an algorithmic form. This does not seem to be acceptable since the cardiologist simply uses quantitative information (in the form of a record) to derive a decision and so it is very likely that his mental process can be imitated (or surpassed) by an algorithm. He may be using more complex relationships between individual features than he thinks and any attempt to translate the process into a program must fail because the presumably correct description is not adequate. If, however, more effective methods were used, it could be that his decisions could be duplicated or improved upon. (Let us stress that even the existing programs perform better than humans in certain important respects: for example algorithmic decisions are certainly more consistent than those made by humans.)

IV. Conclusion

We have seen in the first part of this chapter that the design of algorithms for waveform processing can be approached in a relatively systematic way. In particular, the choice of a classification rule can be guided by the results of pattern recognition. The experience with the existing programs (in particular the ECG programs) shows that computer processing can achieve performance comparable to that of specially trained humans, and at a much higher speed. This has been shown even by programs that do not use the most powerful classification rules. It can be concluded that a more systematic use of the available theory combined with a better definition and understanding of certain outstanding problems will eventually lead to the design of perfectly acceptable programs working both very fast and efficiently. This does not mean that human intervention will not be necessary in any case, but that most of the routine work associated with evaluation of records will be taken over by computers.

Let us note that there are other approaches to waveform processing, in particular the so called linguistic approach (which uses computers to attack the problem from a different point of view). We have not discussed them because they require rather specialized knowledge and are, at the present state of development, not suitable for practical use. The interested reader can consult the book by Meisel (1972) for more information.

V. Summary

Automatic diagnosis of medical signals by digital computers has a long history. A great number of programs for this purpose have been written, particularly for ECG, PCG, BCG, pulse analysis etc. Most of the programs attempt to imitate the routine used by a cardiologist reading the same record. It is shown, in this chapter, that a number of mathematical and statistical methods are available, allowing a much more systematic approach to the design of these programs. Use of these methods should make it possible to develope programs which are more accurate, faster and require less computer storage than the existing ones. These more efficient methods should also speed-up the introduction of automatic diagnosis into practice, and thus contribute to a widespread use of the most advanced techniques in health care delivery. This chapter is an attempt to contribute to this goal by (i) acquainting the medical reader with the technical basis of these methods and, hopefully, convincing him of their superiority, and (ii) trying to attract the technically oriented reader to this important and challenging field.

VI. Appendix 1: A Survey of the Basic Notions of Mathematical Statistics

In order to make it possible to read this chapter even to those who never studied the subject of mathematical statistics we are going to describe, in a very qualitative way, the basic notions necessary for its understanding.

Let us have a large group of samples (for example college students) and a variable (a numerical measure of a certain property) associated with each of them (for example the body weight $w(i)$ – the weight of the i-th student; $i = 1, N$). Let us retrieve a sample without using any specific rule (randomly) and ask what is the value of the associated variable. This question cannot be answered accurately unless we measure the value of the variable associated with the sample. Even without measuring, however, we may be able to give an answer which is not complete but still quantitative. So in the example of the student population and the weight of a student, we may know that for example 70% of students weigh between 63 and 80 kg, or that 84% weigh between 59 and 84 kg. We thus know the 'likelihood' of the value of the variable being in a certain range.

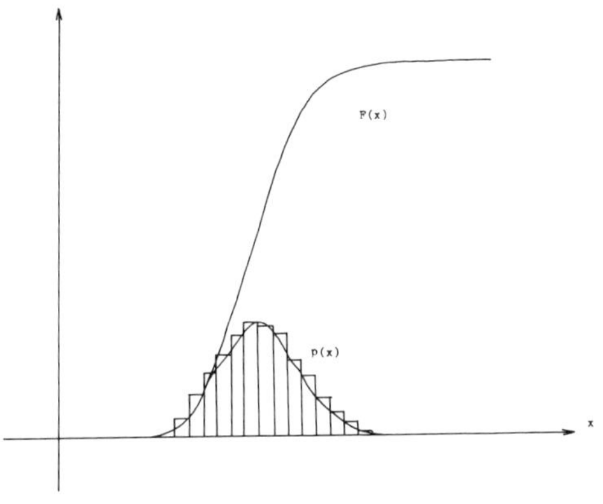

Fig. 27 Two histograms obtained for intervals of different lengths. Probability density as a limit of relative frequencies divided by segment size; distribution function $F(x)$.

This information can be presented more systematically if we construct a histogram (variable value versus the relative frequency of occurence of a value within a specific interval). Let, for example, the minimum and maximum weights be 45 and 95 kg. Let us partition this interval into segments of equal length and plot the relative number of occurences of values within specific segments divided by the length of segment (Fig. 27). When we make the divisions finer, we obtain a smoother histogram (if we have a sufficiently large group). Let us assume that we have available larger and larger groups and that we make finer and finer divisions. If our groups are "representative" (contain variables in "typical" proportions) our diagram will be getting smoother and smoother and approach a limit curve

$$p(w) = \lim_{\Delta \to o} \frac{n(w, w_1)}{N \cdot \Delta} \qquad \Delta = w_1 - w \quad (1)$$

In this expression $n(w, w_1)$ is the number of samples (students) which have variable value (weight) within $|w, w_1|$, N is the total number of samples used. The function $p(w)$ is called a density function of w, and w is called a random variable. This "definition" is meant only as an illustration which shows the basic properties of the notion of a density function. For a rigorous treatment consult any book on mathematical statistics.

The term random variable is derived from the fact that the value associated with a particular sample is unknown before the sample is examined. The density function is the only means which makes it possible to derive quantitative conclusions about w without actually measuring it. There are, of course, other quantitative ways of describing random variables, but they are essentially equivalent and the density function is the most natural choice in the classification problem. A well known equivalent of the density function is the distribution function $F(w)$ defined as

$$F(w) = \int_{-\infty}^{w} p(x)\mathrm{d}x \qquad (2)$$

$F(w)$ can be interpreted as the relative number of samples which have associated variable whose value is not larger than w.

In certain situations we can actually derive the form of $p(w)$ mathematically by the consideration of, for example, the physical meaning of a given problem. This is for example the case in statistical thermodynamics, where assumptions about global rules governing motion of individual molecules can be translated into density functions describing the given phenomenon. In most applications, however, the density function is not known and we have to approximate it in a way similar to that used for approximation of functions given by a few points; we assume that it has a specific form and estimate the unknown parameters from the knowledge of a group of samples (parameter estimation). We have seen that certain classification rules use information of this type, and for this reason parameter estimation is an important problem of pattern recognition.

The number of possible types of density functions is practically unlimited. Any function which satisfies the following two conditions can be a density function:

a) $p(w) \geq 0$ for all values of x; this follows from the interpretation relating $p(w)$ to the relative number of occurences

b) $\int_{-\infty}^{+\infty} r(w)\mathrm{d}w = 1$ (which simply means that if each sample has assigned to it a specific value of w then $w \,\varepsilon (-\infty, +\infty)$).

Of all the possible types of $p(w)$, two are encountered frequently in the literature. There are two reasons for it: these functions can be expected on theoretical and practical grounds to occur relatively often (or be good approximations in many other situations) and they are also easy to handle mathematically. These functions (and the distributions associated with them) are:

a) the uniform distribution

$$p(w) = \begin{cases} \dfrac{1}{b-a} = \text{const for } w \,\varepsilon (a,b) = I \\ 0 \qquad \text{otherwise} \end{cases}$$

$$(-\infty < \mathrm{a} < \mathrm{b} < +\infty). \qquad (3)$$

When w is distributed uniformly in the interval I (Fig. 28) then any value in I is equally likely and other values do not occur at all. This last statement is true only approximately as can be seen from our definition. The density function was defined as a limit, and the limit may be zero even when the terms are not zero; in other words samples outside I can occur but their occurence is extremely rare, practically impossible.

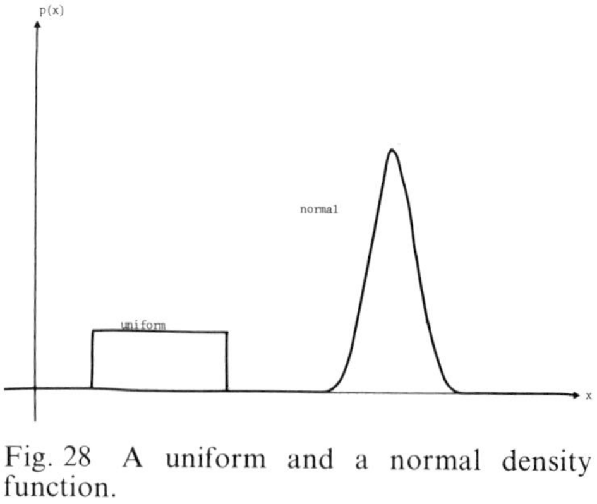

Fig. 28 A uniform and a normal density function.

b) the gaussian (normal) distribution

$$r(w) = \frac{1}{\sqrt{2\pi}\,\sigma} e^{-\frac{1}{2}\left(\frac{x-m}{\sigma}\right)^2} \quad (4)$$

with mean m and variance σ (Fig. 28). Although this function does not look very simple, we have seen in the main text that its use leads to very simple (linear and quadratic) classification rules. It can also been shown that if the value of a random variable is generated under the influence of a large number of factors, its value obeys the normal form under very general conditions. Since this is often approximately the case in practical situations, distributions close to normal are often encountered in practice.

Sometimes instead of a complete description of a distribution (for example by the density function) only a few quantities – the so called statistics – are used. The best known among them is the expected value of the random variable also known as the mean. It is defined as

$$m = E(w) = \int_{-\infty}^{\infty} w\, r(w) \mathrm{d}w \quad (5)$$

We will show that there is a very close relationship between this definition and the usual meaning of this notion. Let us return to our histogram. Let us assume that our segments are narrow enough to make it acceptable to say that the value of the left end of a segment is a good approximation of the value of w for all samples within the segment. Let us assume that our samples have been ordered in such a way that $i > j$ implies $w(i) > w(j)$. Let the value of the left end of the k-th segment be $x(k)$ and the length of the segment $l(k)$. Then the mean value of w can be approximated by

$$\begin{aligned} \bar{m} &= \frac{1}{N}\sum_{i=1}^{N} w(i) = \frac{1}{N}\sum_{k=1}^{K} x(k)n(k) \\ &= \sum_{k=1}^{K} x(k)\frac{n(k)}{N \cdot l(k)} l(k) \\ &= \sum_{k=1}^{K} x(k)p(x(k))l(k) \\ &\int_{-\infty}^{\infty} x\, p(x)\mathrm{d}x = m \end{aligned} \quad (6)$$

($n(k)$ is the number of samples in the k-th interval.)

The larger the size of our group the closer the two values.

Another important statistic is the variance which is defined as

$$\sigma^2 = E[(w - m)^2] \quad (7)$$

It can be shown in a way similar to that used above that variance is closely related to

$$\bar{\sigma}^2 = \frac{1}{N}\sum_{i=1}^{N} (w(i) - \bar{m})^2 \quad (8)$$

(The two parameters of the normal univariate distribution derive their name from the fact that they can be shown to have the meaning defined above.)

In many situations, samples have associated with them random variables but we are not really interested in the particular values of these variables but rather the values of certain (deterministic) functions defined on them. In the same way as we have defined the expected value of a random variable w, we can thus define the expected value of $f(w)$ as that of $y = f(w)$:

$$E(f) = \int_{-\infty}^{\infty} f(x)p(x)\mathrm{d}x \quad (9)$$

This definition can be used to formulate some very useful criteria such as the minimum expected error and minimum expected cost discussed in the main text.

We have given intuitive description of the notions used in this chapter for the case that w is a real number (a one dimensional variable). All of our results can be extended to the general multidimensional (multivariate) case. In this case, w is an n-dimensional vector whose components are real numbers

$$w = (w_1, w_2, \ldots w_n) \quad (10)$$

As a particularly important and non-trivial extension, let us mention the multivariate normal density function

$$p(w) = \frac{1}{\sqrt{(2\pi)^n}\sqrt{D}} e^{-\frac{1}{2}(w - M)^T \Sigma^{-1}(w - M)} \quad (11)$$

Here n is the dimension of w, Σ is the $n*n$ covariance matrix (with a meaning similar to that of the variance in the one-dimensional case), M is the vector of mean values and D is the value of the determinant of Σ.

A more rigorous survey of the most important statistical terms used in pattern recognition can be found in Fukunaga's book on statistical pattern recognition (1972) or any book on statistics (read, for example the book by Ehrenfeld and Littauer (1964)).

VII. Appendix 2: Automatic Conversion of Signals

The purpose of this appendix is to acquaint the reader with the basic notions of conversion of continuous signals to digital form. References for more detailed study are given in the last paragraph.

Signals recorded from the human body usually have the form of a voltage continuously changing with respect to time. In other words, at any moment the recorded voltage has a certain value and this value is continuously changing in time. Recording of a signal of this type on paper results in a continuous line. Signals of this type are usually called *analog signals*.

Digital computers, as their name indicates, work with digital values, i.e. discrete entities. This means that an analog signal is not a meaningful datum for a digital computer. To represent a signal for computer processing, signal values have to be sampled at discrete time intervals (Fig. 29) and the measured voltages converted to a sum of a fixed number of terms, powers of a base (usually 2). So for example, number 11 can be represented as

$$11 = 8 + 2 + 1 = 1*2^3 + 0*2^2 + 1*2^1 + 1*2^0 \quad (1)$$

This can be expressed simply as 1011. (Note that the usual decimal notation uses the same principle.) If this form is represented by a physical variable, such as a voltage, the resulting signal is called a *digital signal.*

Waveforms recorded on paper can be converted into digital form by hand. When it is necessary to convert large amounts of data this method would obviously be very laborious. In these cases, automatic conversion by analog to digital (A/D) convertors is necessary. An A/D convertor accepts continuous analog voltage and outputs discrete digital signals suitable for computer processing. The user only has to decide what parameters the A/D convertor should have on the basis of the nature of his signal and the requirements of his problem.

In certain applications, mostly the control of industrial processes by digital computers, we have to perform the inverse conversion, from the digital into the analog form. This is done by digital to analog (D/A) convertors. D/A convertors are very important in A/D conversion as well, since the conversion of an analog signal to a digital one is usually performed by comparison of the current value of the analog signal with a discrete voltage generated by the convertor. This voltage is usually obtained as a combination of basic voltage steps by means of switches controled by logical circuits. The value of the generated voltage which is closest to the analog voltage is retained and the corresponding digital value is output as its digital approximation.

A/D conversion has several important effects upon the nature of the resulting digital approximation:

1) Since each of the values handled by the computer is stored in one location of computer's memory, only values at discrete points are retained from the original continuous signal. The signal is thus sampled (usually at regular

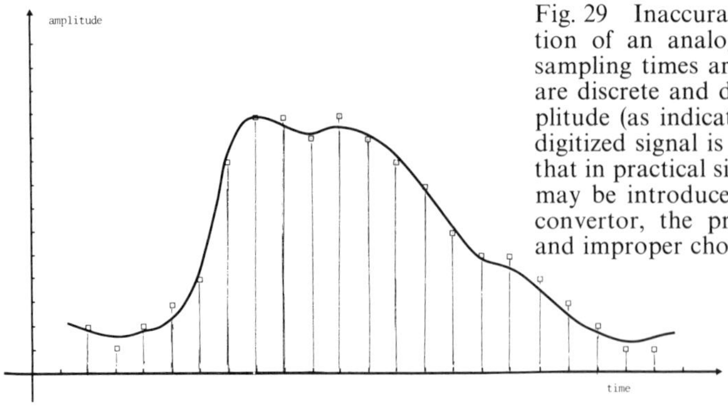

Fig. 29 Inaccuracies introduced by digitization of an analog signal: a) only values at sampling times are available, b) digital values are discrete and differ by steps of a fixed amplitude (as indicated on the axis). The shown digitized signal is an ideal output in the sense that in practical situations further inaccuracies may be introduced by the design of the A/D convertor, the properties of its components and improper choice of sampling rate.

intervals), converted and the digital approximation stored. Graphically this means that instead of a continuous line only values at certain points are available for further processing (Fig. 29).

2) Since each number is expressed as a sum of a fixed number of terms (bits), the value of the signal cannot be approximated with unlimited accuracy (in the example given above the number of bits was 4 and the arrangement described would not make it possible to approximate number 11.342 by a value closer than 11). The stored number is thus only an approximation of the true value (Fig. 29).

3) A/D convertors, as any other devices, have limited accuracy and their operation thus adds more or less to the error described under 2.

4) A/D convertors are limited in the maximum number of samples which they can convert in a unit of time (the sampling rate). The shortest time required by a convertor to read an analog signal and convert it into a digital form is in the order of microseconds.

5) It was indicated that conversion is performed essentially by comparison of the analog signal with a voltage obtained as a combination of voltages supplied by open or closed switches. Since each switch corresponds to one bit in the resulting digital signal, comparison is in fact performed bit by bit; adding an extra bit (increasing the accuracy of approximation) thus increases the number of operations the convertor has to perform. This implies that if an analog signal is to be converted with greater accuracy (requiring more bits), the maximum sampling rate decreases since more operations have to be finished before the next sample can be processed.

It follows from the enumeration above that some of the important considerations in the choice of parameters of an A/D convertor are:

a) The accuracy of approximation, obtained by dividing the range of signal values by the number of units, must not be worse than the minimum required precision. If, for example, the input to the convertor is in the range of (0, 1) volts and the accuracy must be at least 0.0005 volts, then at least 10 bits are required to achieve it. This is because in order to obtain approximation with the maximum difference between the actual value and the approximation not greater than 0.0005, the difference between two successive digital values must not exceed 0.001 volts. For example, to approximate 0.5554 volts with accuracy at least 0.0005 it is enough to have available steps 0.555 and 0.556 (and 0.555 is the required approximation). It is therefore neccessary to have at least 1000 digital steps. 10 bits guarantee more than that since $2^{10} = 1024$; 9 bits are not enough since $2^9 = 512$.

b) Another factor is the desired sampling rate: It was shown that for the same type of convertor the maximum available sampling rate decreases with increasing number of bits. Since we often use one convertor to sample several analog signals in succession (for example several leads of the ECG can be sampled in a fixed order one by one), the maximum available sampling rate may not really be comfortably large. The minimum sampling rate suitable for a given application depends upon the nature of the signal and the highest frequency useful component we want to retrieve undistorted: If the highest frequency is f, we have to take, theoretically, at least $2*f$ samples/second; this condition is known as the sampling theorem. If, for example $f = 200$ in each ECG lead, we have to take at least 400 samples/second in each lead. If, for a given number of bits, the required sampling rate is too large, we have to make a compromise: We may have to accept somewhat lower accuracy (using fewer bits to store sampled signals), and/or slower sampling, sacrificing part of the high frequency components of the analog signal.

The presentation of A/D conversion in this appendix is only illustrative and does not exhaust all considerations that a prospective user of an A/D convertor has to make. For additional information the Digital Computer User's Handbook (Klerer, Korn 1967) contains the basic facts presented in more detail. A detailed technical description can be found, for example, in Schmid (1970).

Glossary of Technical Terms

A priori probability: In our context the probability that an arbitrary sample belongs to a particular class

Algorithm: A step by step description of a process (in our context usually a numerical procedure) easilly convertable to a mechanical implementation (a computer program)

Analog signal: A signal whose value is the magnitude of a physical variable (e.g. electrical voltage), continuously defined with respect to time and usually continuously changing

Artifact: In our context an undesirable component of a signal whose source is usually other than that of interest—e.g. spurious electrical noise affecting the recorded signal

Bayes rule: See minimum expected error rule

Conversion: Transformation of signals from analog into digital form or vice versa

Classification of a sample: Assignement of a sample into one of several classes (e.g. diagnosis of an ECG record as a normal or abnormal one)

Clustering of samples: Grouping of samples on the basis of their similarity (e.g. clustering of complexes in an ECG record)

Cost of classification: A numerical value associated with classification through costs associated with correct or incorrect assignement of a sample

Decision rule: A rule which assigns a sample to a class on the basis of the numerical value of the associated feature vector

Design set: Set of labeled samples used to derive a decision rule

Density function: A statistical description of the distribution of random samples in the sample space

Digital signal: A quantized signal, whose value can change only in steps

Euclidean distance: The common square root distance

Expected value: Essentially the mean value of a function defined on random variables, calculated over a large number of samples

Fast Fourier transform (FFT): A computationally efficient method for converting signals from the time domain representation into the frequency domain representation and vice versa

Feature recognition: (automatic) Identification of the feature vector on a given sample (e.g. identification of the Q,R,S points in an ECG record)

Feature selection: Selection of representative values which describe the signal for the purpose of classification

Feature vector: A vector describing the sample to be classified (e.g. the coordinates of the Q,R,S points of an ECG complex). Its components are numerical values of individual features.

Frequency domain description: Description of a signal in terms of amplitudes of sine and cosine components, the sum of which can be used to represent the signal in the time domain

Labeled sample: A sample whose classification is known (as opposed to a sample whose classification is to be determined by a decision rule)

Linear and quadratic discriminant functions: Linear and quadratic functions of values of feature vectors. Their value is used to classify samples

Linear filtering: Numerical or physical (usually electronic) elimination or attenuation of certain undesirable components of the signal

Minimum expected error rule: An optimal decision rule whose application results in classification with the minimum proportion of errors when used on a large number of samples

Minimum expected risk rule: An optimal decision rule whose application results in classification with the minimum cost, when used on a large number of samples

n-dimensional space: Set of all n-dimensional vectors

n-dimensional vector: The set of n numerical values (e.g. if an ECG complex is described by the amplitude of the Q,R,S points, it is represented by a 3-dimensional vector whose components are the corresponding amplitudes)

Nearest neighbor rules: Family of decision rules in which assignement of a sample is determined by the classification of its (labeled) nearest neighbors from the design set

Normal (Gaussian) distribution: An often encountered type of statistical distribution in which the density function has an exponential form

Optimal decision rule: One which minimizes a prescribed cost of classification

Parameter estimation: Calculation of approximate values of parameters describing in our case statistical distributions

Piecewise-linear approximation: Approximation of a signal by a succession of straight lines

Preprocessing: Preparation of a signal before classification to make possible application of a decision rule (e.g. linear filtering, feature recognition)

Random variable: In our context a numerical value associated with a given random sample

Reject option: Class of decision rules which reject (do not assign) samples which could not be classified with sufficient certainty

Representative: In our context a feature vector used to represent a class

Sample: In our case a biomedical signal representing a subject (e.g. an ECG record)

Sample space: The set of all possible samples. Each sample in an n-dimensional sample space is represented by an n-dimensional feature vector

Statistical pattern recognition: A theoretical discipline dealing mainly with the problem of classification of samples. It is based on mathematical statistics

Test set: Set of labeled samples used to test a decision rule

Time domain representation: Description of a signal by its values measured at consecutive instants of time

Uniform distribution: One for which the corresponding density function is either 0 or a positive constant in every point of the sample space.

References

1. Alexander D C: Multichannel simultaneous analysis of the electrocardiogram using simultaneously recorded 3 lead sets. In: Clinical electrocardiography and computers (Caceres C A, Dreifus L S, eds). Academic Press (1970)
2. Caceres C A, Dreifus L S: Clinical electrocardiography and computers. Academic Press (1970)
3. Cady L D, et al: A method for ecg wave pattern estimation. Circ Res 9, 1078–1082 (1961)
4. Chow C K: On optimum recognition error and reject tradeoff. IEEE-IT 16, 41–46 (1970)
5. Cox J R, et al: Digital analysis of the electroencephalogram, blood pressure wave and the electrocardiogram. Proc IEEE 60, 1023–1031 (1972)
6. Duda R C, Hart P E: Pattern classification and scene analysis. John Wiley L Sons (1973)
7. Ehrenfeld S, Littauer S B: Introduction to statistical method. Mc Graw-Hill (1964)
8. Fukunaga K: Introduction to statistical pattern recognition. Academic Press (1972)
9. Fukunaga K, Narendra P M: A branch and bound algorithm for computing k-nearest neighbors. IEEE Computer Society Repository R 74-122 (1974)
10. Gold B, Rader Ch M: Digital processing of signals. Mc Graw-Hill (1969)
11. Hart P E: The condensed nearest neighbor rule. IEEE-IT 14, 515–516 (1968)
12. Klerer M, Korn G A: Digital computer user's handbook. Mc Graw-Hill (1967)
13. Klingeman J D, Pipberger H V: Time-shared computer analysis of orthogonal electrocardiograms. In: Clinical electrocardiography and computers (Caceres C A, Dreifus L S, eds). Academic Press (1970)
14. Litwin W, Begon F: Digital processing of phonocardiograms: first results. Journal of Bio-Medical Computing 5, 59–65 (1974)
15. Marriot M R, et al: Review of the present state of automated electrocardiogram interpretation. CMA Journal 108, 1230–1232 (1973)
16. Meisel W S: Computer-oriented approaches to pattern recognition. Academic Press (1972)
17. Moore D H: Combining linear and quadratic discriminants. Comput Biomed Res 6, 422–429 (1973)
18. Mucciardi A N, Gose E E: A comparison of seven techniques for choosing subsets of pattern recognition properties. IEEE-c- 20, 1023–1031 (1970)
19. Pavlidis T: Linguistic analysis of waveforms. In: Software engineering vol 2 (Tou J, ed). Academic Press (1967)
20. Perry L W, et al: Computer analysis of the phonocardiogram. In: Engineering in the practice of medicine (Segal B E, Kilpatrick P E, eds). Williams & Wilkins Company (1967)
21. Pipberger H V: Computer analysis of electrocardiograms. In: Clinical electrocardiography and computers (Caceres C A, Dreifus L S, eds). Academic Press (1970)
22. Previtt J M S: A versatile clustering algorithm with objective function and objective measure. Comp Programs Biomed 2, 297–314 (1972)
23. Schmid H: Electronic analog/digital conversions. Van Nostrand Reinhold (1970)
24. Smith R J: Circuits, devices, and systems. John Wiley & Sons (1971)
25. Specht D F: Generation of polynomial discriminant functions for pattern recognition. IEEE-EC- 16, 309–319 (1967a)
26. Specht D F: Vectorcardiographic diagnosis using the polynomial discriminant method of pattern recognition. IEEE-BME- 14, 90–95 (1967b)
27. Stallmann F W, Pipberger H V: Automatic recognition of electrocardiographic waves by digital computer. Circ Res 9, 1138 (1961)
28. Tomek I: A clustering technique in the classification of almost periodic waveforms. Comput Biomed Res 5, 621–628 (1972)
29. Tomek I: Two quantitive measures of similarity of one class of waveforms. Comput Biomed Res 6, 499–508 (1973)
30. Tomek I: The iterative nearest neighbor – a preliminary report. Conference 74, CIPS/DIPS. Ottawa, May 9–12 (1974a)
31. Tomek I: Piecewise-linear approximation with a bound on absolute error. Comput Biomed Res 7, 64–70 (1974b)
32. Tomek I: More on piecewise-linear approximation. Comput Biomed Res 8, 568–572 (1975)

33. Tomek I: Generalization of the *K-NN* rule. IEEE-SMC- 121–126 (1976 a)
34. Tomek I: An experiment with the edited nearest neighbor rule. IEEE-SMC- 448–452 (1976 b)
35. Tomek I: A comparison of two methods of feature identification of waveforms. (Not published. 1976 c)
36. Weihrer A L, et al: Computer programs for an automatic electrocardiographic system. In: Clinical Electrocardiography and Computers (Caceres C A, Dreifus L S, eds). Academic Press (1970)
37. Wilson D L: Asymptotic properties of nearest neighbor rules using edited data. IEEE-SMC- 408–421 (1972)
38. Young T Y, Calvert T W: Classification, Estimation and Pattern Recognition. American Elsevier Publishing Company (1974)

I. Tomek,
Department of Computer Science,
ACADIA University,
Wolfville, BOP IXD, Nova Scotia
(Canada)